IEE POWER AND ENERGY SERIES 37

Series Editors: Professor A. T. Johns
D. F. Warne

ELECTRICAL STEELS

for rotating machines

Other volumes in this series:

ELECTRICAL STEELS

for rotating machines

Philip Beckley

The Institution of Electrical Engineers

Published by: The Institution of Electrical Engineers, London,
United Kingdom

The Institution of Electrical Engineers,
Michael Faraday House,
Six Hills Way, Stevenage,
Herts. SG1 2AY, United Kingdom

British Library Cataloguing in Publication Data

Beckley, P. (Philip)
Electrical steels: for rotating machines −(IEE power and
energy series; no. 37)
1. Steel, Electrical
I. Title II. Institution of Electrical Engineers
620.1′1297

ISBN 0 85296 980 5

Typeset by Newgen Imaging Pte, India
Printed in Great Britain by Bell & Bain Ltd., Glasgow

Dedication

To Mary Beckley, who supported the computer work.

Contents

Foreword

This book is intended to offer to the electrical design engineer an insight into what properties may be expected from electrical steels and how these properties may best be exploited in machine design. It is hoped that by giving insight into the interplay between magnetic properties, manufacturing methods, physical properties and costs, sound rational judgements may be made about machine structuring and operation.

No attempt is made to delve into the inner recesses of the theories of physical metallurgy; suffice it to make clear what does in fact happen for what clearly traceable reasons.

It is hoped that by offering this book to electrical engineers, their connections with material producers will be more productive to the benefit of all.

Acknowledgements

Grateful acknowledgement is made to COGENT POWER Ltd. (formerly European Electrical Steels), for permission to use an extensive range of material data, and a variety of pictures and diagrams drawn from its own publications.

Also thanks go to:

The UK Patent Office, Figure 3.3
The USA Patent Office, Figure 3.3
The former British Iron and Steel Research Association, Figure 3.6
School Science Review, June 1960, Figure 3.8c
The American Society of Materials, Figure 3.10
Editors of the Laugton-Warne Electrical Engineers Handbook, Table 6.1
The Wolfson Centre, Cardiff, Figures in Chapter 9
Instron Ltd., Figure 10.56
Wolfson Centre, Figure 12.2B
British Standards Institute, Figures 14.1 and 14.2

Thanks are also due to David Rodger and Faris Al-Naemi for their modern insights into finite element work in Chapter 9.

Chapter 1

Introduction

One of the deep mysteries of the Universe is action at a distance. Gravitation, electric and magnetic forces operate across empty space and through all sorts of material.

Gravity is so much a familiar part of everyday life that it took philosophers a long time to realise that there existed a force to be considered. Indeed the concept of a force and its action on matter was slow to emerge in a quantifiable form.

Electric forces were evident in the ancient world as attractions due to electrostatic surface charges and sparks from the movement of dry clothing. Amber and silk have been known since antiquity and in dry weather offer observable electrical effects.

Very little was to be seen of magnetism in ancient times. The aurorae gave interesting displays in northern regions but little else to provoke thoughts of magnetism. Tradition has it that shepherds in part of ancient Turkey (Magnesia) found that their iron shod staffs became attracted to certain rocks (Figure 1.1). The regional name became attached to the effect and 'magnetism' entered the world vocabulary.

Figure 1.1   Shepherds experience a magnetic force

Figure 1.2   Natural lodestone

Eventually these effects became exploitable, as pieces of these natural rock magnets showed directional properties when hung up by a thread. A magnet suspended in this way would swing round to point in a north–south direction. So the lodestone (leading stone) evolved as a primitive navigational compass (Figure 1.2). It was observed that steel needles rubbed by a natural magnet themselves became magnetised and formed a much more convenient compass component. A needle could be combined with a piece of cork to float on water. A loose pivot stabilised the movement of needles.

From such primitive beginnings the art of magnetism expanded into a science leading to the studies of Galileo (1564–1642) and Gilbert (1544–1603).

It may be supposed that natural magnetite rocks became magnetised through the influence of lightning strikes. Millions of ampères flow into the Earth over a small area during a lightning stroke. Statistical treatment of the arising magnetising fields and the worldwide pattern of lightning strikes suggest that the chance of a magnetite rock remaining unmagnetised is small. As engineers know well, once a magnetisable material becomes a permanent magnet the probability of getting it back to a zero-flux unmagnetised state by later random field applications is small.

Experiments with primitive magnets led to the idea of magnetic poles. Of these, like poles repel each other and unlike attract (Figure 1.3). A 'north seeking' pole is the end of a magnet that points to the Earth's north magnetic pole when freely suspended.

The development of primary cells and the production of relatively steady currents of electricity (Galvani and Volta) prepared the ground for Oersted to observe that a

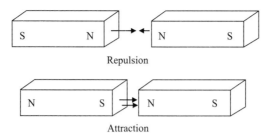

Figure 1.3   Like poles repel each other, unlike attract

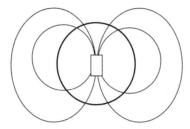

Figure 1.4   The magnetic field surrounding the Earth is such as might be produced by a short bar magnet buried within it

current flowing in a conductor would influence a compass needle. The magnitude and direction of this influence is related to the size and direction of the current involved. Very soon electromagnets were built in which coiling up the current-carrying conductor concentrated the magnetic influence.

It fell to Faraday, in a series of magnificent experiments, to show that time-varying magnetism would induce current in a nearby electric circuit, and then to show that time-varying current would induce electromotive forces in a separate conductor system. Faraday's laws expressed the electromagnetic continuum with which we are familiar today.

Faraday's laws may be set out as:

1. When a conductor moves with respect to a magnetic field, an emf is developed in that conductor.
2. The magnitude of the emf developed is proportional to the rate of change of the magnetic flux.

It should also be noted that Lenz's law requires that the polarity of the field generated by induced currents is such as to oppose the motion producing it. Were this not so perpetual-motion machines would become possible and the laws of thermodynamics would fail.

Within this set of discoveries is the implication that if a current-carrying conductor can act as a magnet, might not metal or oxide magnets contain currents which give the same effect? Further, may not the Earth itself contain a giant magnet to produce the terrestrial magnetism so useful for navigation (Figure 1.4). In fact the liquid metal core of the Earth is considered to give rise to a dynamo action producing the terrestrial field. The precise mechanism of this is obscure and a variety of explanatory theories are extant. It can be shown that the polarity of the Earth's field has reversed many times in pre-history. This is evident from magnetism 'frozen-in' to the rocks of ancient hearths. Chaos theory may be able to explain the erratic behaviour of the Earth's field on an appropriate time-scale.

1.1   The habits of iron

Iron and steel can be induced to show a range of magnetic effects. Hard steel once magnetised tends to remain so. Soft iron is easily magnetised but its magnetised state disappears when the field inducing it is removed. A piece of 'soft' iron shows no overt sign of magnetism and will not alone pick up pins or behave in a magnetic way.

The great breakthrough in electromagnetism was the discovery that electric currents produced magnetic fields and that changing magnetic fields, when linked with a conductor, produced electric effects. The extension of this experience is that two electric current systems each producing a magnetic field will react on each other and produce interactive forces just like 'permanent' magnets do.

The 'discovery' of the electron gave a picture of electric currents being considered in terms of electron flow. 'Discovery' really means the assembly of a set of theories which cohere well and have strong predictive power. This led to the question, could electrons be active inside an iron bar to produce magnetic effects? Detailed research into the atomic structure of iron and other metals showed that in fact it is the electron spins within an iron crystal lattice that deliver magnetic effects as normally observed.

Further, electron spins would need to be spatially organised if powerful external effects were to be observed. This is in fact what happens, but not to many elements. Studies by Bethe showed that only when atomic parameters fall into a certain

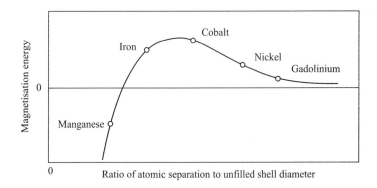

Figure 1.5   Bethe's curve

range will spontaneous co-operative alignment of electron spins become energetically favourable (Figure 1.5). Thermodynamic experience indicates that whenever a lower energy state is available Nature will preferentially occupy it.

Iron, cobalt and nickel are the prime candidates for so-called 'ferromagnetism'. Non-ferromagnetic materials exhibit a weak response to magnetising forces, either reinforcing or opposing them. Such paramagnetic and diamagnetic effects are important to physical theories but do not yield forces of engineering importance.

1.2 Ferromagnetic domains

Accepting that the crystal lattice of iron is such that electron spins spontaneously align to give a powerful overall directionally co-ordinated magnetic field, why is this not evident to the outside world when magnetic effects are sought in the vicinity of an iron bar? Representing iron as having a set of spontaneously coupled electron spins, it ought to look like Figure 1.6. These coupled spins would give a powerful magnet. It ought to have north and south seeking poles as it interacts with the Earth's field. The well-remembered scattering of iron filings should plot a field round it as shown in Figure 1.7. So why not?

1.3 Energy storage

It is a universal principle of physics than whenever a lower energy condition can be found Nature will tend towards it. (Pace 'are we humans a low energy state, and if not why are we here?') The picture we have of a strong internal spontaneous magnetic field aligning the electron spins has consequences for energy storage. Alignment of all the spins leads to a distortion of the iron lattice: it becomes a bit longer in the direction of overall magnetisation, an effect known as magnetostriction, which we will look at as a practical effect later. All the aligned spins in iron necessarily come

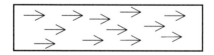

Figure 1.6   Spontaneously coupled electron spins in iron

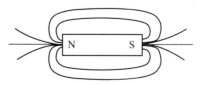

Figure 1.7   Magnetic field from iron

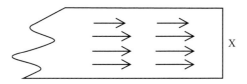

Figure 1.8   End of a magnet

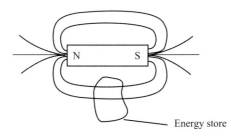

Figure 1.9   Energy stored in a field

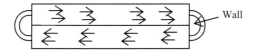

Figure 1.10   Domain wall

to a stop at the boundary of the metal. The point X (Figure 1.8) will experience a magnetic field there in the ways we are familiar with. The extension of the internal field outside the metal into space can be represented as energy storage in that space (Figure 1.9), an energy which Nature is keen to minimise if it can.

A rearrangement of spins to produce such an effect looks like Figure 1.10. Depending on the theory employed, the energy could be considered as residing in the magnet in terms of effort needed to keep the spins aligned when those near the end next to space lack aligned neighbours, but anyhow there is stored energy. What then about this sort of situation?

If the aligned spins comprised two sets of opposite direction, the stored energy, overall, is much smaller with a lower external field. This subdivision can be continued, situations (2) and (3) in Figure 1.11 having progressively lower stored energies and smaller external fields. Looking at (2) we could imagine parts A and B as separate domains of magnetism pointing in opposite directions. If continued to condition 4, no field emerges and pins are never picked up. We now have the situation of the iron being powerfully magnetised (inside each domain), yet with the domains so organised that the effect outside the metal is negligible (the no-pin-pick-up situation).

What happens now if such a domain-based metal is exposed to an outside magnetic field? The region A in Figure 1.12 will be reinforced and tend to grow and the

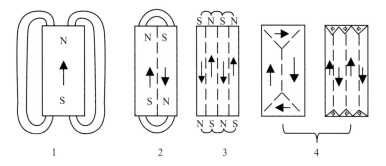

Figure 1.11   Progressive reduction of magnetostatic energy

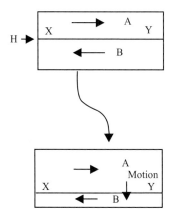

Figure 1.12   Movement of a domain wall

region B will be opposed and tend to shrink. The so-called 'domain wall' XY will move sideways to make this happen. A rather feeble applied field will command the movement of the domain wall XY so that the effective observable magnetism coming from the metal will rise from zero to the high value associated with full alignment of electron spins over all the iron lattice. Reversing the sign of the applied field leads to a reversal of the overall magnetisation. The magnetisation produced by this process of domain wall movement is very powerful, and can increasingly be up to a million times that which would have resulted from the same magnetising force acting on air or empty space.

This simple view of domains as the 'concealment system' of ferromagnetism, and the motion of their walls being the route of expression of ferromagnetism in the engineering world, is far from a complete picture but it will suffice at this stage of our discussion.

Within 'soft' iron the ferromagnetic state of magnetisation to saturation via electron spin coupling hides in the pattern that domains adopt for energy minimisation.

In later chapters we will look in more detail at the features which control domain size and shape, how motion of their walls leads to minute changes in the physical size of steel components (magnetostriction) and how purposeful immovability of walls gives us 'permanent magnets' in 'hard' materials.

The unique property of ferromagnets lies in the ability of comparatively small externally applied fields to control the development of large amounts of magnetic flux which can involve itself in electromagnetic activities. When the crystal structure of a material makes such a response to applied fields easy and reversible, the material is referred to as magnetically 'soft' (which often goes hand in hand with mechanical softness). When the crystal structure is organised so that domain wall motion and magnetic response requires a great applied field to produce it, and when the

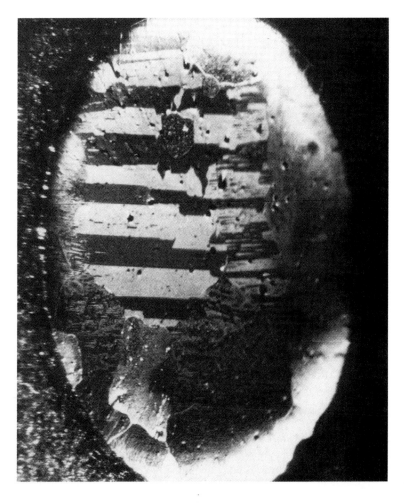

Figure 1.13   Domain structure revealed by magneto-optic imaging

realised magnetisation remains largely in place when an external field is removed, the material is referred to as 'hard'. 'Permanent' magnets of this type have their uses in electrotechnology, but are, as a class, distinct from electrical steels which are designed to be soft. Techniques have been developed to make the internal magnetic domain structure visible and this is further described in the literature [1.1] [1.2]. Figure 1.13 shows the ferromagnetic domain structure observed in a sample of electrical steel.

The electrotechnical world is, today, well supplied with 'amplifiers' which govern energy flow and power deployment in response to tiny signals. Mechanical amplifiers (servomechanisms) have long been used and the controllable flux magnifying abilities of ferromagnets has given rise to magnetic amplifiers of various sorts during the twentieth century. It is however the exploitation of flux magnification in transformers and rotating machines which has made the electrotechnology of the modern world possible.

A class of materials exist – ferroelectrics – whose properties mirror those of ferromagnets. They have not so far offered an equivalent exploitability in power engineering.

References

1.1. BRAILSFORD, F.: 'Physical principles of magnetism' (Van Nostrand, London, 1966).
1.2. CAREY, R. and ISAAC, D.: 'Magnetic domains' (EUP, London, 1966).

Chapter 2

The functions of electrical steel

2.1 General

There are two main applications of electrical steels: in transformers and rotating machines. Transformers benefit from the flux-magnifying properties of ferromagnets by enabling high rates of flux change to be produced in a small compass so that efficient transformers of manageable size can be built. We will not consider transformers further in this book, but further reading is readily available [2.1] [2.2] [2.3] [2.4].

2.2 Force magnification

The force operating between current-carrying systems (magnets) can be shown to relate to $B^2 A$ when A is the area under consideration and B is the magnetic flux density over that area. Thus in any machine the force which can be produced depends on B, and B depends on how much flux can be produced from a ferromagnetic material.

Iron can be processed so that it will magnify by thousands-fold the magnetic field acting upon it. When fed through to B^2 the effect on force can be million-fold. When force moves through a distance work is done and the rate of force movement is expressed as power in horsepower, watts or kW. So the use of a ferromagnetic material to magnify flux leads directly to enormous magnification of the power potential of machines of a given size and weight. The production of magnetic fields from flowing current involves power loss in ohmic resistance (except for superconductors in some circumstances), so that the field-magnifying properties of iron saves on materials as well as energy.

Figures 2.1, 2.2 and 2.3 show the effect of an applied magnetic field on a so called magnetic material over a wide range of fields. It can be seen that ferromagnets such as iron represent only a small anomaly on the cosmic scale, and it is within this 'small' bubble of anomaly that all electrotechnology has its existence.

Figure 2.4 shows the response of steel to an applied magnetising field compared with air. At first gains are small, then a huge surge of flux becomes available, but

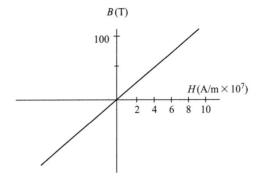

Figure 2.1   B v. H looks like a simple straight line for high fields

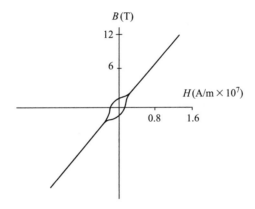

Figure 2.2   An anomaly is visible near the origin

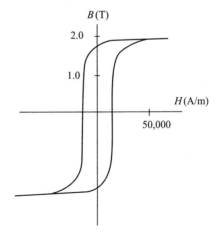

Figure 2.3   At commonly experienced fields the familiar B–H loop emerges

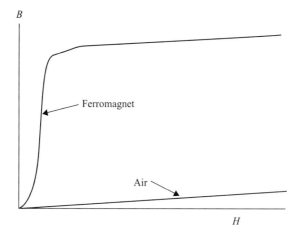

Figure 2.4 Comparative response of air and a ferromagnet

beyond about 2 tesla the yield peters out and the iron-filled space is little better than empty space so far as flux production is concerned. The 2 tesla produced at little 'cost' in terms of applied magnetising field may be considered as 'free'.

When feeble fields are used to control large amounts of flux which can relate to large mechanical forces, the system has the properties of an amplifier with positive gain. Magnetic amplifiers can take the form of static or rotating devices, but for specifically power amplification such devices are almost obsolete.

2.3 Units

In the early days of magnetic studies forces between magnets were used to describe their behaviour and this led to a system of units involving magnetic poles and the like. Since that time the unit system employed in magnetism has been revised many times. We will concern ourselves with modern units only.

2.4 Stimulus–response

Essentially the whole of magnetism centres round the idea of a stimulus applied and a response engendered (Figure 2.5). The stimulus is called a magnetising field and is denoted by the letter H for its quantity. The response of a material is denoted by the letter B. Modern units are now all electrically based and will be considered as follows.

2.4.1 Magnetising field, H

This is measured in amps/metre and relates to the number of ampères flowing round (for example) a solenoidal region (Figure 2.6). The conductors carrying the ampères

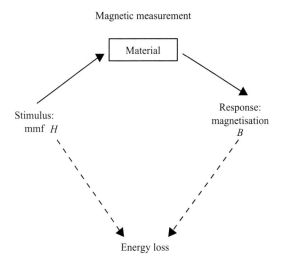

Figure 2.5   Magnetisation as stimulus and response

round may be large or small, few or numerous; only the aggregate number of amps flowing per metre length of enclosure counts.

A distinction needs to be made between magnetising field and magnetising force. A magnetising field experienced in a given region (e.g. within a solenoid as in Figure 2.6) has an intensity defined as that which would be produced within a long solenoid around which an aggregate current of H amps/metre was flowing. By contrast, a magnetomotive force is represented by the total magnetising 'effort' put into a magnetic circuit independent of its spatial layout.

Consider a ring (Figure 2.7) of soft magnetic material. For each ampere (I) fed to the coil winding C, n amp turns of magnetomotive force are applied to the circuit. The coil could be tightly wound, or spread out. If wound over a length l the amp turns/metre operating within the wound region are $I \times n/l$, but the magnetomotive force is $\int_0^l H \, dl = n \times I$. Magnetomotive force may be provided by a current-carrying winding or a permanent magnet of equivalent effect.

It is convenient to use a solenoidal form as a starting point. The development of this from Ampère's law relating to straight conductors can be found in many textbooks. Older technology used the oersted as the unit of applied field, and this is still found, even in current work on relay steels, permanent magnets, etc., especially in the USA. 1 oersted = 79.58 A/m.

2.4.2   *Magnetisation or magnetic induction, B*

This is measured in tesla, which is a unit describing the quantity of magnetic flux per cross-sectional area, leading to the weber. The weber is defined as the amount of magnetic flux which, if removed or introduced steadily into a region bounded by a conductor in 1 second, will induce an emf of 1 volt in the conductor during the

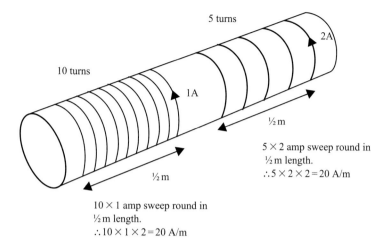

10 turns

5 turns

1A

2A

½ m

½ m

5 × 2 amp sweep round in
½ m length.
∴ 5 × 2 × 2 = 20 A/m

10 × 1 amp sweep round in
½ m length.
∴ 10 × 1 × 2 = 20 A/m

Figure 2.6   Notion of amps/metre

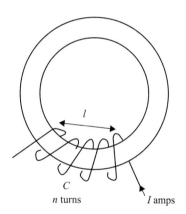

l

C

n turns

I amps

Figure 2.7   mmf delivered by a winding

activity, that is: 1 weber = tesla × area. The Weber is seen to describe absolute amounts and the tesla the area intensity of magnetisation.

The degree of response to stimulus is thought of as the *permeability* of the material concerned, that is, the level of B induced by an applied field H such that permeability = B/H. μ is used as the symbol for permeability. Free space is, of course, one simple medium and has a permeability of $4\pi \times 10^{-7}$, so that writing $B = \mu_0 H$, for free space $B_{tesla} = 4\pi \times 10^{-7} \cdot H_{a/m} = \mu_0 H$.

It may be wondered what property permits the existence of a magnetic field in empty space. Whether the air in front of my face (very like free space in terms of magnetism) sustains a magnetic field or not, what property of that space sustains it? This is a philosophical question harking back to the luminiferous aether of long ago

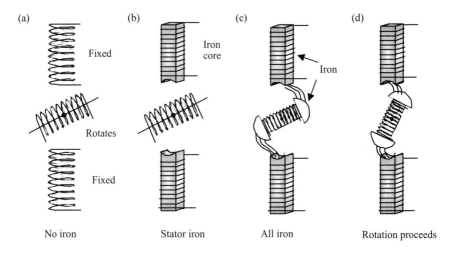

Figure 2.8   Flux magnification by the use of ferromagnetic coress

and points towards theories of the zero-point energy of space in modern cosmology. Further, does the propagation of magnetic fields, e.g. over cosmic distances, require a quantised system of magnetons, etc?

When considering ferromagnetic materials of usefully high permeability the μ in $\mu = B/H$ involves both the underlying response of free space to H, as well as the 'extra' due to electron spin alignment and domain activity, so that $B = \mu_0 H + M$, M being the extra magnetisation induced. Its units too are tesla.[1] Sometimes J is used rather than M. Since H is in A/m, and B in webers/m^2, and webers in volts seconds, analysis of the units will show that the units of permeability are henries/metre. (Recall a henry is the inductance which will deliver 1 volt of back-emf when the current in it changes at a rate of 1 ampère/second.)

A ferromagnetic material, then, exploits the spontaneous self-magnetisation by internal electron spin alignment and the tendency of this spontaneous magnetisation to be hidden by a self-cancelling domain structure. The ability of a small applied field to reorganise the domain structures and make co-ordinated electron spin effects for the whole sample available to the outside world in a controllable fashion lies at the core of magnetic usefulness of steel as a component of machines.

Magnification of flux leads via B^2A to force magnification. This could be thought of in terms of machines with progressively more iron used in their construction (Figure 2.8). To give scale to the units involved, 10,000 amps flowing in a region 1 m wide is quite a large current; 100,000 is becoming an engineering problem. Iron saturates at a little over 2 T and can have a relative permeability in the thousands.

---

[1] Older texts may use gauss rather than tesla (1 T $=$ 10,000 gauss). North American texts may use lines/inch^2 where 1 line/inch^2 $=$ 0.1555 gauss/cm^2.

References

2.1. CONNELLY, F. C.: 'Transformers' (Pitman, London, 1950; later reprints, e.g. 1962).
2.2. COTTON, H.: 'Applied electricity' (Cleaver Hume, London, 1951). Various books by Cotton cover the subject in greater depth.
2.3. HEATHCOTE, M.: 'J and P transformer book' (Newnes, Oxford, 1998).
2.4. KARSAI, K., KERENYI, D. and KISS, L.: 'Large power transformers' (Elsevier, Amsterdam, 1987).

Chapter 3

History of the development of electrical steels

The discovery that an iron core would greatly enhance the power of a lifting magnet led to the use of ferromagnetic materials being incorporated in the moving and stationary parts of motors and generators. In the earliest times, cast iron and wrought iron were used since they were convenient and available. The permeability of cast iron is not very good, as seen in Figure 3.1.

The use of solid cores in machines causes heavy penalties in the form of loss-making eddy currents so that the products of the developing sheet metal industries became incorporated into machine cores. A solid iron core is an electrical conductor and acts as a short-circuited turn in which strong currents are induced as magnetisation changes. By making the core using a pile of thin sheets or a bundle of wires the eddy currents are much diminished.

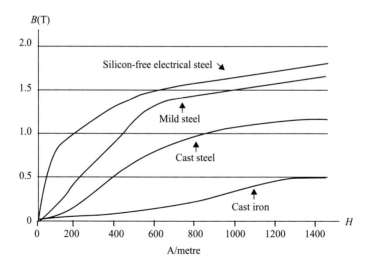

Figure 3.1   Magnetisability of materials by moderate applied fields

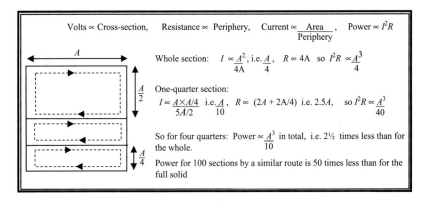

*Figure 3.2   Lamination restrains eddy currents and reduces power wastage –
a simplified model*

Power losses were still high and even if power efficiency was not of great concern, certainly the tendency of machines to run too hot was troublesome. Around the turn of the 19/20th century it was found that raising the resistivity of iron by alloying with silicon greatly helped to restrain the flow of eddy currents. Figure 3.2 shows the benefits of eddy currents being denied free range in solid cores; large reductions in power loss are possible.

Early patents on alloying with silicon were filed by Barret, Brown and Hadfield. Increasing electrical resistivity by alloying reduces the intensity of eddy currents. Figure 3.3 shows the areas of technology covered. Early versions of electrical steels were produced by hot rolling using hand-fed mills processing sheets which were usually rolled in packs. Hot-rolled electrical steel made by this method was very isotropic in properties, that is, its magnetic and physical properties were almost the same in the rolling and 90 degrees to rolling directions, and it was possible to handle silicon contents up to some 4.5 per cent. Above 4.5 per cent silicon brittleness becomes unacceptable. A limited amount of cold rolling of sheets also took place by rolling individual sheets.

The advent of strip cold rolling, pioneered in the USA, opened the gate for high-speed strip rolling in which coils of steel thousands of metres long were processed in continuous tandem-stand or reversing mills. The particular properties of cold-rolled steel have vastly influenced the directions in which electrical steels have developed. Silicon levels are restricted to about 3–3.5% due to rolling behaviour.

At room temperature, iron has a body centred cubic lattice and within a crystal of iron in which electron spins have spontaneously aligned to give self-saturation, some directions are much more easily magnetisable than others. Figure 3.4 shows a representation of the iron lattice and plots the relative magnetisability of various directions. When iron crystallises the individual crystallites are usually randomly arranged so that the magnetic properties of the overall material represents an average of the various crystal directions.

RESERVE COPY

Nº 3737

A.D. 1902

Date of Application, 13th Feb., 1902
Complete Specification Left, 15th Dec., 1902—Accepted, 15th Jan., 1903

PROVISIONAL SPECIFICATION.

Improvements in Electrical and Magnetic Apparatus such as Transformers, Dynamos, and other Appliances, and Alloys for use therein.

I, ROBERT ABBOTT HADFIELD, of Parkhead House, Sheffield, in the County of York, Steel Manufacturer, do hereby declare the nature of this invention to be as follows:—

In the construction of electrical and magnetic apparatus such as transformers,
5 dynamos, motors and other appliances where a direct or alternating current is employed to generate magnetic lines of force, iron or mild steel has been used, involving considerable loss owing to the inadequate permeability of the iron or steel and the consequent considerable hysteresis loss or dissipation of energy in each magnetic change or reversal.
10 Now I have discovered that the efficiency of apparatus such as referred to may be greatly increased by the use in the construction thereof of alloys containing

---

No. 745,829.

Patented December 1, 1903.

UNITED STATES PATENT OFFICE.

ROBERT A. HADFIELD, OF SHEFFIELD, ENGLAND.

MAGNETIC COMPOSITION AND METHOD OF MAKING SAME.

SPECIFICATION forming part of Letters Patent No. 745,829, dated December 1, 1903.

Application filed June 12, 1903. Serial No. 161,228. (No specimens.)

To all whom it may concern:

Be it known that I, ROBERT ABBOTT HADFIELD, a subject of the King of Great Britain, and a resident of Sheffield, county of York,
5 England, have invented certain new and useful Improvements in Magnetic Compositions and Methods of Making the Same, of which the following is a specification.

My invention relates to material having
10 magnetic and electrical properties suitable for use in various electrical apparatus, such as ballast-coils, transformer-plates, and the like.

The object of my invention is to produce
15 an improved material of this character having specially high permeability and electrical resistance and low hysteresis qualities. I have found that material of these desirable qualities can be produced by alloying iron with
20 other elements, among which I will name silicon and aluminium, phosphorus also yielding satisfactory results, as well as combinations of two or three of these elements.

I may proceed, for instance, as follows: I
25 take pure Swedish or other suitable pure iron

and low hysteresis for efficient use in transformers and other electrical apparatus in which said qualities are useful.

I have found that the superior qualities of
55 my improved material of alloy can be still further enhanced by a treatment involving alternate heating and cooling and generally carried out as follows: I first heat the material to between about 900° and 1,100° centi-
60 grade and allow it to cool, preferably quickly. Then I reheat the material to between about 700° and 850° centigrade—that is, to a temperature lower than the one attained during the first heating—and then allow the metal to cool
65 very slowly. In practice the cooling has been often extended to last several days. Either one or both of these treatments may be frequently repeated, or after the first treatment has been carried out the second type of heat-
70 ing may be frequently repeated. I have, for instance, taken a steel alloy of the composition above mentioned, heated it to 1,070° centigrade, cooled it quickly to atmospheric temperature, reheated it to 750° centigrade,
75 cooled slowly, again reheated to 800° centi-

Figure 3.3   Front section of the Hadfield patents

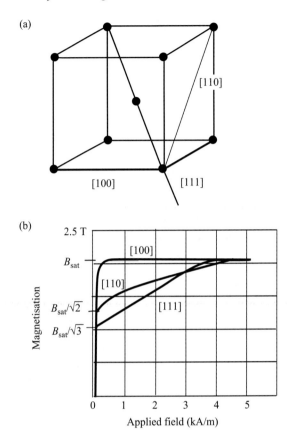

Figure 3.4 *a* Body centred cubic lattice: [111] is a body diagonal; [110] is a face diagonal; [100] is a cube edge

  b The [100] direction is the easy cube edge direction; [110] is the hard cube face diagonal direction; [111] is the hardest cube body diagonal direction

Cold rolling leads eventually to crystallites which are not quite randomly arranged, so that the aggregated magnetisability varies somewhat in directions other than that direction referred to as the rolling direction. This will be looked at quantitatively later.

By the application of a complex series of metallurgical rolling and heat treatment operations, grains of metal can be caused to grow such that an 'easy' direction of magnetisation lies along the strip rolling direction. Such a material has greatly enhanced magnetic properties in the rolling direction, but relatively poor ones in the (transverse) 90° to rolling direction. Such properties are unwelcome in most rotating machines but excel in transformers where unidirectional flux is desired.

Some very large alternators can use stator cores built up of grain-oriented steel with appropriate attention to exploitation of directionality.

3.1 Tools to improve magnetic efficiency

The identifiable stages of material improvement are listed below. They all have associated costs so that the material applied to end use is that which has the appropriate property/cost trade-off.

3.1.1 Lamination

Eddy currents are restrained by lamination as shown in Figure 3.2. Iron is quite a good conductor of electricity so that if solid iron is being used as a magnetically permeable core the outer surface of the iron could be viewed as a low-resistance short-circuit enclosing the core. This skin will have emfs induced in it during magnetisation reversals and the consequential 'eddy currents' waste energy by dissipating heat. If the core metal is subdivided into thin sheets the balance of eddy current path resistance and induced emf shifts so that the overall power wastage in the core is radically reduced.

By the operation of Lenz's law eddy currents produce magnetic fields tending to oppose the field changes producing them. The effect of this is to delay the penetration of magnetisation arising from applied fields from reaching the centre of the core. This means that the core metal is in effect under-used and will not fully contribute to the circuit magnetic flux if the frequency of reversal is too rapid.

Restraint of eddy currents by lamination and raising of resistivity promotes rapid flux penetration and full usage of the iron. It does of course incur the costs of making thin steel and insulating one layer from the next. Further, if overdone, very thin laminations lead to a poor percentage space occupancy by iron rather than by surface inhomogenieties. Also, surfaces act as prime domain pinning sites and some loss components rise if steel is too thin. All in all the range 0.2–1.0 mm covers most applications.

3.1.2 Raised resistivity

The addition of silicon raises resistivity. The degree of this influence is shown in Figure 3.5. Aluminium also raises resistivity usefully, but there are difficulties associated with its use. Aluminium is oxygen-avid and this can emerge as alumina inclusions which pin domain walls.

Enormous effort has been applied to attempts to discover binary, tertiary or quaternary alloys able to perform better than silicon as resistivity raisers. Figure 3.6 (after Foley) shows the outcome of an enormous project on this subject. Remarkably the silicon used in 1903 remains the overall most successful alloying element. A process similar in effect to raising resistivity can be produced by breaking up metal into tiny particles insulated from each other and embedded in an electrically non-conductive matrix. This is the so-called soft magnetic composite route and may be expected to be important for operating at high frequencies.

Adding silicon has disadvantages. These are:

(i)   Cost of ferrosilicon to be added to iron: Silicon is available commercially in the form of ferrosilicon, and ferrosilicon of specially low carbon content is more expensive but helps to keep the carbon level in the final steel sheet low.

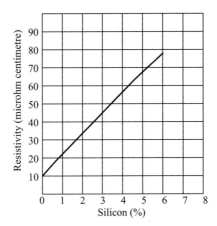

Figure 3.5   Relationship between electrical resistivity and silicon content

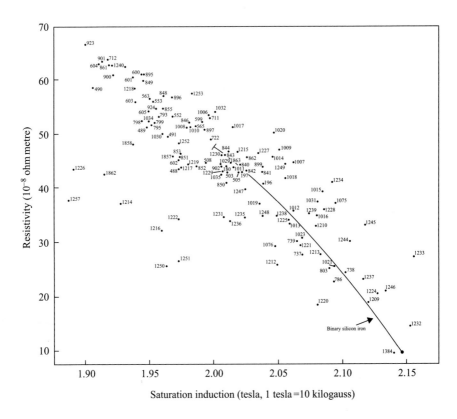

Figure 3.6   Effect of alloying elements on resistivity and saturation magnetisation [after Foley et al. BISRA]

(ii) Greatly increased brittleness induced by added silicon: Alloying elements often affect the physical properties of iron when present in quite 'small' quantities. Carbon in the range 0–1% covers a whole range of important steel properties, and for electrical steels carbon as low as 0.002% is desirable. Silicon up to some 3–3.5% can be added while maintaining fair rollability.

(iii) Decrease in the total flux available from iron due to dilution by silicon: Remembering that mechanical power is proportional to B^2, losses in B due to dilution are soon felt.

However, the presence of silicon reduces some of the ill-effects of impurities as will be seen.

The ability of hot rolling to process 4.5% silicon steel is offset by the fact that the surface of hot-rolled steel is rougher than cold-rolled, leading to a reduced space factor (see later). Cold-rolled steel is smoother, but beyond about 3.25% silicon ductility under cold rolling is inadequate.

When silicon dissolves in iron it does so without creating any second phase of crystals which may be non-magnetic or act as a stress site or domain wall obstruction in the range of interest for electrical steels.

3.1.3   Purification

Any non-metallic inclusions in steel impair its magnetic properties. This arises because when magnetisation is being changed domain walls must move, and they are pinned and obstructed by non-metallic inclusions. A domain wall carries energy implicit in its existence and if a 'hole' is produced in a wall due to its straddling an inclusion (Figure 3.7) then this will constitute an energy well as energy is needed to create wall area as the wall is pulled away. Figure 3.8(a) and (b) shows this and Figure 3.8(c) plots the force versus distance acting on a wall. Consequently it is important to carefully restrain the presence of elements which produce non-metallic inclusions. Prime offenders are carbon and sulphur. It is desirable to keep carbon and sulphur at a minimum during steel-making (see later on very low carbon steel). Carbon dissolves quite readily in hot iron so that an annealing treatment takes any small amount of carbon into solution. Unfortunately this solution is not stable and

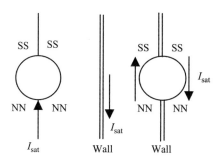

Figure 3.7   Pinning of a domain wall on an inclusion

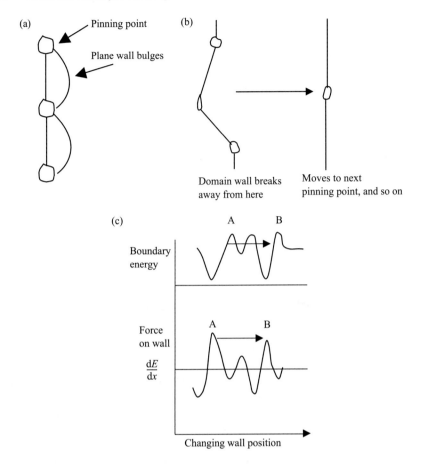

Figure 3.8   *a*   Displacement of pinned walls
              b   Wall jumps from one pinned position to the next
              c   Energy and force plots for changing wall position

carbon will precipitate in the form of non-magnetic carbides if the metal is operated at, for example, 100 °C for many hours.

When a domain wall pulls away from an inclusion by way of a sudden wall movement, wall velocity is momentarily very high. Rapidly moving walls induce micro-eddy-currents in the metal lattice and these are more intense at higher wall velocities. Micro-eddy-currents lead to extra joule (eddy current) heating and raised overall losses.

At a macroscopic level carbide precipitates which 'hang up' domain walls increase the coercive force (relating to the ease of demagnetisation) of steel and increase the area of hysteresis loops. The extra losses due to the effects of precipitation of carbides raise the temperature further so that more precipitates form, and so on. Early machines suffered from this 'ageing' effect and could require periodic re-annealing of cores.

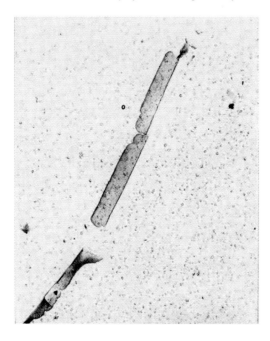

Figure 3.9   Ageing precipitate ×10,000

Chemical steps can be taken to restrain precipitation, e.g. addition of titanium, but if overdone the effect is worse than the ailment.

An alternative strategy is 'overageing'. In this practice it is possible to hold metal for a long time at a moderate temperature where precipitation happens and particles grow till the precipitated material exists as rather few larger species. These produce less overall *domain wall hold ups* per % carbon. Very tiny precipitates produce stress fields around them in the lattice and stressed regions also pin domain walls. Figure 3.9 shows a micrograph of an ageing precipitate. Sulphur is less tractable than carbon and is best removed early in steel-making. This removal is expensive.

3.1.4   Grow large grains

When steel is held at a high temperature, e.g. 800–1000 °C, grains grow at the expense of each other so that overall grain size increases. With large grains the presence of the grain boundary is less per unit volume. Grain boundaries are prime domain wall pinning sites so that large grains reduce the overall amount of obstruction of wall movement per unit volume. There are various technologies for promoting the rapid and extensive growth of grains which will be considered under manufacturing methods.

3.1.5   Orientation of grains

We have seen that the iron lattice has directions (cube edges) which are specially favourable for ease of magnetisation and this is exploited in the case of transformer

Figure 3.10   Dr Norman P. Goss

steel. Here the so-called Goss orientation is used. Norman P. Goss (Figure 3.10) developed grain orientation processing in the 1930s in the USA after finding that suitably applied cold rolling and heat treatment regimes led to the selective growth of grains having their easy directions of magnetisation in the strip rolling direction. For use in most rotating machines Goss orientation is not suitable as it does not have isotropic properties in the plane of the sheet. It is possible to consider grain orientations which are more favourable than random yet suitable for use in rotating machines.

If the random cube-on-face pattern (Figure 3.11) is considered it is clear that there are many cube edge directions in the plane of the sheet, as well as face diagonals, but the least favourable body diagonals are excluded. The production of a well-defined random cube-on-face texture (a crystal arrangement is spoken of as 'texture') would give a good result for motor steel but has yet to be fully realised by metallurgists at an economic price.

While it has been suggested that large grains are a 'good thing' it can be overdone. For various thermodynamic reasons it is found that the domain wall spacing increases as grain size increases. This means that if domain wall spacing is large, then individual walls must move at high speed to get the magnetisation change done in the available time. A rapidly moving wall is a generator of micro-eddy-currents associated with the magnetisation vector rotation happening within it, so that high-speed walls produce

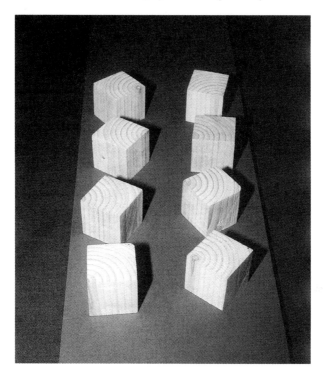

Figure 3.11   Random 'cube-on-face' texture, suitable for rotating machines. There are no body diagonals in the plane of the sheet

extra losses. There will therefore be a grain size which, if exceeded, is unfavourable (maybe 5 mm across in transformer steel which grows very large grains as part of the preferential orientation process) so that special techniques have to be used to artificially control domain wall spacing. Until recently it was considered that this situation does not arise in steel for motors/generators and 'grains-as-big-as-possible' remained an unqualified aim for such steels. However the rapid rise in the use of motor speed control using pulse width modulation (PWM) and allied methods has led to dB/dt values which call for wall motion so rapid that consideration will have to be given to creating an optimised grain size.

3.1.6   Removal of stress

Stressed regions in steel constitute domain wall pinning sites and are to be avoided, so that stress-relieving heat treatments are often used to optimise magnetic properties. In general stress, and particularly compressive stress, impairs magnetic properties. The strained displacement of atoms in stressed regions affects the crystallite response to magnetisation and efforts to remove stress are well rewarded. The usual method of stress removal is the application of an annealing treatment when atoms can relax to equilibrium positions. However, for best results annealing should be the last treatment applied. If punching of laminations is last the benefits of annealing are partly lost. Further, if motor stator stacks are forced into shell housings with great force the machine core is in a permanently stressed state and cannot perform optimally.

3.1.7   Tensile coatings

Not all stress is unfavourable and for some steels (e.g. transformer steels) coatings which induce tension in the underlying strip can be helpful. This approach is usually confined to grain-oriented transformer steels.

3.2   Reviewing the tools available

Lamination
Alloying to raise resistivity
Purification
Large grains
Oriented grains
Stress relief
Tensile coatings.

The progressive development of the above techniques has led to improvements in magnetic quality which have accrued over the last 100 years. In the next section an examination of how these tools are deployed will be undertaken.

Chapter 4

Manufacturing methods

4.1 Iron-making

Steel derives from iron-making which was once wholly derived from ore smelted in the conventional blast furnace. More recently direct reduction processes are coming into use in which a less capital-intensive practice applies. In each case iron ore (oxide) is reduced to iron metal. Impurities become incorporated into crude iron. Further, the use of scrap steel to feed the steel-making process is a growing practice. Of course scrap does not come into existence unless metal is won from ore at some earlier time, but the effects of scrap usage must be considered.

Iron is mined all over the world and taken to iron-making sites far from the quarrying region. It is possible then to blend ores to yield particular iron compositions. As-smelted blast furnace iron contains some 4% of carbon and varying amounts of manganese, phosphorous, sulphur, silicon and other trace elements. Steel-making amounts to removal of the non-ferrous elements and readjusting the composition by additives to give steels best fitted to various purposes.

Much of the carbon is removed at steel-making. Some elements are desired in iron, others not. Small amounts of silicon help with oxygen removal, and the balance of manganese and sulphur can influence the grain orienting capabilities of steel. Phosphorous affects final hardness. Titanium is an undesired element but removal is not practicable, so that low-titanium ores are desired for making iron destined for electrical steels. Heavy use of scrap can entrain more copper (from scrap electrical devices) than is best for final steel properties.

A typical blast furnace iron may be as set out in Figure 4.1. Steel-making of the ordinary sort, 'mild steel', leads to compositions as shown in Figure 4.2. By the manipulation of liquid slags present at steel-making some elements can be removed or their amounts altered. Formally an ordinary mild steel may be of some 0.1% carbon + traces of other elements.

```
Carbon 4%
Manganese 1–2%
Phosphorus 0.4–1%
Sulphur 0.05% max.
```

Figure 4.1   Iron composition

```
Carbon 0.1%
Manganese 1%
Phosphorus 0.05%
Sulphur 0.05% max.
```

Figure 4.2   Ordinary mild steel composition

For electrical steels special emphasis is placed on post-steel-making treatments aimed at:

(a)   Precision additives:
Si for resistivity
P for hardness control
Al for some types of grain orientation
Mn helps resistivity.
(b)   Removal so far as practicable:
C by special treatments
S by special treatments
N_2 by avoiding pick-up
O_2 by avoiding pick-up.

Over the past two decades the removal of carbon and sulphur down to very low levels, e.g. less than 0.003%, has become possible at the secondary steel-making stage. This has altered the scope of later processing where the carbon levels had to be separately addressed during strip production. Carbon can be removed down to very low levels by a process called 'vacuum degassing' in which liquid steel is exposed to high vacuum and agitation (Figure 4.3).

Sulphur is removable by treatment of the liquid metal with calcium silicide. All these treatments are expensive but can pay dividends in avoiding the need for expensive treatments during the post-rolling stages of production.

4.2   Outline of steel processing

Figure 4.4 shows the outline of a steel strip process route. Liquid steel is continuously cast into slabs which are hot rolled down to some 2–4 mm thick coil. This coil is called 'hot band'. Hot rolling to thicknesses less than 2 mm is problematical as accurate

(a)

(b)

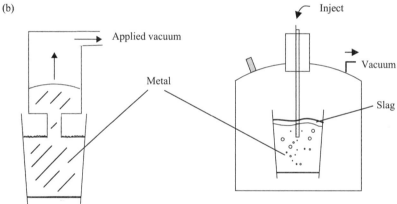

Figure 4.3   *a*   Vacuum degassing installation
            b   Vacuum degassing units

thickness control and good surface condition are harder to maintain so treatment is continued cold. A few hot mills can approach 0.7–1.0 mm for hot-rolled final gauge (thickness), but this is not common.

Modern cold mills take in hot-band coils and roll them progressively thinner to a final thickness of 0.2–0.7 mm (Figure 4.5). Before cold rolling, coils will be exposed to a shot-blast, side-trim and pickling process to remove scale and provide a surface suitable for cold rolling. A strand anneal may precede first cold rolling to homogenise the internal structure of the metal. In a strand anneal a coil of steel is fed into a long furnace as it is unrolled, then coiled up again at the far end. Furnaces may be hundreds of metres long.

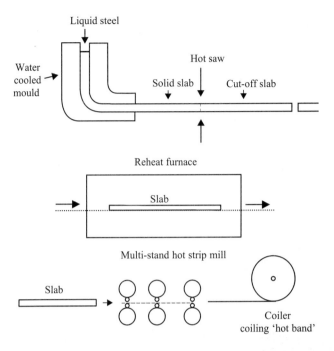

Figure 4.4   Steel casting and slabbing

4.2.1   Decarburisation

When steel is rolled to thin final thicknesses such as 0.5 mm or less it is possible to decarburise it by exposure to a decarburising atmosphere as the strip passes through a strand annealing furnace. To obtain effective decarburisation, exposure to wet hydrogen at 800+ °C for a minute or two is effective. Carbon diffuses to the surface and reacts with the furnace atmosphere gas to produce hydrocarbons, carbon monoxide and carbon dioxide.

   If decarburisation is not performed during steel-making and not carried out during a later strand anneal treatment, it must be tolerated or removed from the finally punched lamination before use. A considerable proportion of steel for motors is annealed after punching to remove internal stress and the effects of a final cold reduction. Remember that internal stress inhibits easy domain motion. This final anneal will be carried out at a motor builders plant or a lamination supply house which offers finished laminations to the market. Decarburisation at this final stage of heat treatment requires a long cycle at 800+ °C set up to ensure that decarburising gases can penetrate to every part of a stack of laminations in the furnace load and that the products of reaction have time to diffuse out. Additionally, heat must reach all parts of the furnace charge so that metal is hot enough for long enough. Figure 4.6 outlines the strand decarburising process, in this case one in which a final oxide coating is being applied as part of a grain orientation process.

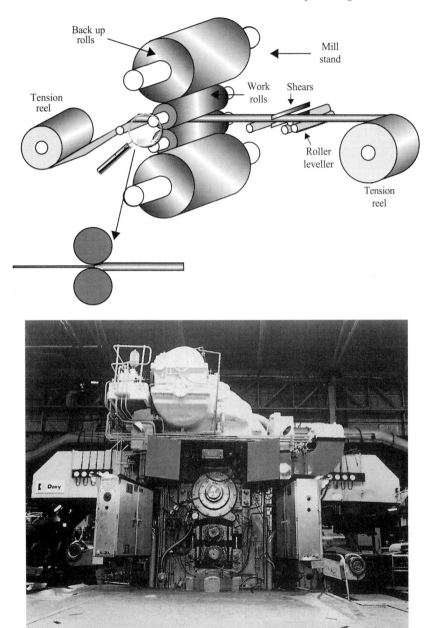

Figure 4.5   Cold rolling mill, schematic and photograph

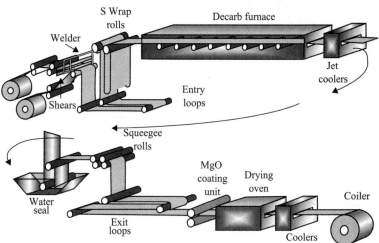

Figure 4.6   Photo and schematic of a decarburising line

In the process shown below (iron to final product) decarburising can be applied at stages 3, 5 or 7, or the effort spread over the three processes:

1. Iron-making
2. Primary steel-making
3. Secondary steel-making
4. Cold rolling
5. Strand annealing
6. Final cold reduction
7. Post-stamping decarburising anneal.

Depending on the duty for which steel is required and the excellence of magnetic performance desired, decarburisation can be applied with varying degrees of effort (and cost).

Long ago decarburisation was achieved by reaction of the metal when hot with air (oxygen). This formed a surface oxide scale in which carbon was incorporated and which could be pickled away. Of course iron was lost in the scale also, so it was never a popular process.

Early attempts to reduce carbon levels before the cold-rolling stage involved winding coils of hot band (as-hot-rolled some 3 mm thick) steel with an interlap wire mesh and heating them in a decarburising atmosphere using a batch process. The interlap mesh permitted gas penetration between the laps of a coil, and the annealing treatment could be continued long enough to allow good carbon removal. Speed of removal of carbon varies approximately as the inverse square of steel thickness so that a long box treatment of coils was made. This 'open coil annealing' treatment had some merit, but the provision and recovery of an interlap mesh was a troublesome process. It may be noted that steel very low in carbon and other hardening elements is very soft and of low mechanical 'Q' so that a piece is readily bent by hand and if dropped emits a thud rather than a clang.

4.2.2   Desulphurisation

Sulphur levels have traditionally been kept low by management of slags during steel-making so that sulphur is encouraged to migrate into supernatent liquid slag and be removed. Much lower levels of sulphur (e.g. less than 0.002%) can be attained by treatment at the secondary steel-making stage with calcium silicide or other compounds. The cost of this treatment is considerable, but very low-sulphur metal, when of low carbon content also, gives excellent magnetic performance.

4.3   Cold-rolling methods

The input to cold-rolling mills is normally 'hot band' steel some 2–4 mm thick and approximately 1 m wide. Coils are typically of 20 tonne weight. Figure 4.7 depicts such a coil. A rolling mill could be thought of, crudely, as an oversize mangle in which strip to be reduced passes through a pair of rolls which are compressed onto it

Figure 4.7   Hot rolled coil – hot band

as shown in Figures 4.8(a) and (b). To secure considerable reduction of thin strip the physics of the roll bite requires rolls to be of comparatively small diameter. Small-diameter rolls are more susceptible to bending under the loads applied to them, so that the centre region of strip being rolled can be reduced in thickness less than the edges. To counter this effect back-up rolls are used to restrain deformation of the 'work' rolls (Figure 4.8c).

More complex arrays of multiback-up roll systems are used in special Zendsimir mills used for very heavy reductions. Other expedients involve rolls with non-parallel profiles and the facility of internal inflation of rolls by hydraulics.

Rolling, both hot and cold, is a very extensive technology and its ramifications need not concern steel users.

Mills may be tandem or reversing. In the tandem mill steel is progressively reduced in thickness roll-stand to roll-stand whereas in a reversing mill the metal passes from left to right through one mill stand, then back again, through several 'passes' to attain the final desired thickness.

The surface condition of hot-rolled steel is never as smooth as that attainable by cold rolling. The achievement of the cold (reducing) mill is assessed in terms of:

- Speed of working
- Exactitude of output thickness (gauge)

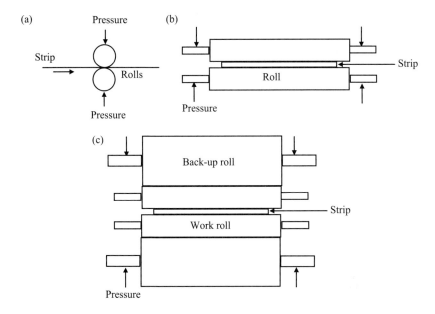

Figure 4.8   *a*   Outline of a rolling mill
              b   Pressure applied to roll necks
              c   Use of back-up rolls to stiffen a rolling mill

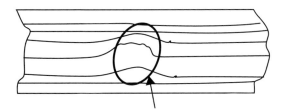

Figure 4.9   *Example of shape aberration: loose middle with tight edge*

- Perfection of 'shape' (see Chapter 10)
- Cleanliness
- Strip profile
- Surface roughness.

The output thickness is important to the motor builder who expects to have a clear appreciation of how many laminations will produce a stator stack of specified height. Chapter 13 examines the steps taken to procure exact thickness control.

'Shape' describes the state of deformation or internal stress of the strip. The rolling process can operate so that the filaments of metal nearer to the edge of the strip become less longitudinally extended than the centre regions (Figure 4.9). This state is known

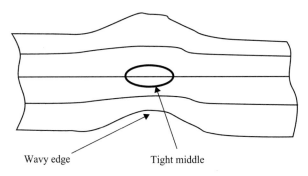

Wavy edge   Tight middle

Figure 4.10   Example of shape aberration: tight middle, wavy edge

as loose middle or tight edge. Similarly the edge regions can be over-extended leading to a wavy edge (Figure 4.10). These 'shape' variations are often invisible when strip is under tension, and are present in the form of differential internal stress.

Punching strip of unfavourable shape into laminations can produce defects such as round holes (as punched) becoming oval when relaxed. Such an effect is clearly bad for motor stator laminations where a close-tolerance round hole is vital. Fortunately effective methods of continuous shape control can now be applied on mills. Output shape can be sensed at various positions across the width of the strip and signals applied to a range of corrective mechanisms.

Cleanliness: Clean strip requires clean rolling, clean lubricant and clean ingoing strip. Chapter 10 looks at cleanliness assessment.

Surface roughness: It might be thought that the more perfectly smooth a steel's surface could be the better it would be. However, too smooth a surface can provoke sticking together (by micro-welds) of laminations during stack annealing processes, and will inhibit the penetration of gas between laminations during a final anneal when access for gas to produce decarburisation may be desired. To control surface roughness final cold rolling may be carried out using 'textured' rolls which have been precision roughened to give a desirable surface roughness to steel strip.

Strip profile considers the variation of thickness of strip across its width. There are two main components of such variation – crown and edge drop. Crown refers to the overall growth of thickness towards the centre of the strip, and edge drop considers the more rapid fall away of thickness in the last 100 mm near the strip edge. For motor laminations the variation in thickness side-to-side of a lamination is a factor of first importance. If edge drop is severe, sloping stack profiles can arise (Figure 4.11).

The progressive rotation of laminations as they are stamped to even out stack shape is a useful procedure but is less widely practised than would be most useful. Of course by discarding some edge material the effect of edge drop can be minimised, but this is a very expensive option. Strip profile is much more easily controlled during hot rolling than during cold rolling, so that specificational liaison between hot and cold mills is essential. The provision of near-perfect profile is expensive since rolls wear in use and despite many compensational ploys will require regrinding earlier if

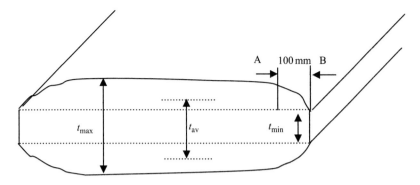

Crown = $t_{max} - t_{av}$. 'Edge drop' is the thickness change over the edge region A–B. The thickness variations are shown on a greatly exaggerated scale.

Pile of laminations

A lamination pile cut from steel with taper will not pile level unless individual laminations are rotated.

Figure 4.11   Edge drop and crown

best profile is continuously demanded. It should be noted that metal of perfect 'shape' can have severe crown or edge drop and vice versa. The two problems have differing origins and different solutions.

4.4   Heat treatments

A wide range of heat treatments can be applied to steel. These may be aimed at softening steel between sequential parts of a cold-rolling regime (interannealing), at producing homogenisation of dissolved elements/compounds, or at promoting grain growth.

It is beyond the scope of this book to consider the details of the physical metallurgy of low-alloy steels, but it may be noted that the equilibrium phase diagrams of iron with elements such as carbon and silicon show that phase changes may or may not occur as material is heated and cooled through critical temperatures. Grains may re-grow during phase change so that careful management of heat treatment is needed to optimise grain size and form. Even though equilibrium phase diagrams may be well charted, actual processing has to occur at a finite speed so deviation from equilibrium results may be expected. High skill and extensive insight is required to achieve the desired process results reliably.

4.4.1 Effects of final anneal

When a final anneal after stamping is also a decarburising anneal performed on steel where vacuum degassing has not been used to produce ultra-low carbon, the wet hydrogen atmosphere used can lead to some subsurface oxidation of the steel which damages magnetic permeability. Thus the final anneal of ultra-low carbon steel should be carried out in a dry or neutral atmosphere, e.g. 97% dry nitrogen + 3% hydrogen. Hydrogen mops up small amounts of leaked-in oxygen from the air.

Unfortunately plant set up to perform a decarburising anneal may also have to be used to process ultra-low carbon steel and in doing this the very best potential properties of ultra-low carbon steel can be impaired. Here a balance between simplicity and adjustment costs has to be found.

It may be pointed out that in popular traditional terms 'steel' is an alloy of carbon in iron, so that the term 'ultra-low carbon steel' is oxymoronic. However the core metal based on iron is referred, throughout manufacturing, as 'electrical steel' exact definitions notwithstanding.

4.5 Critical grain growth

It has been noted that large grains help to reduce power loss (less grain boundary present). An optimised grain size may be attained by two main routes. The application of these involves careful consideration of their respective costs and of the relevance of the coatings applied to steel.

4.5.1 Method 1 – long, hot, soft

In this approach steel is final strand annealed so as to give it a lengthy exposure to temperatures ranging up to 1000 °C for long enough to allow grains to grow considerably. After this treatment steel may be cooled and coated, either in the same production line or at a separate coating unit. If this steel is then punched into laminations it is large grained and annealed and thus ready for use. However the punching operation will have introduced shearing stress to the laminations and in tooth areas of stator laminations the ratio of bulk/near-cut-edge material may make this state unfavourable. However the limited detriment of shearing stress may be accepted rather than applying a further final anneal. Organic coatings which most facilitate punching do not respond well to an 800 °C final anneal so acceptance of shearing stress may be a reasonable option. The punching of fully soft steel can provide a challenge to stamping operators if they are steeped in a culture of punching hard steel (see below).

4.5.2 Method 2 – critical grain growth

Many stamping operators favour the use of 'hard' steel as punching feedstock. What is 'hard' steel? A VPN of 140–180 is considered by stampers to be crisp and tractable. Material of 90–120 VPN is considered to be 'soft'. (VPN = Vickers Pyramid Number, a scale of hardness widely used in the strip steel world.)

If soft steel is applied to presses designed to punch crisp steel, dragging and smearing of metal in the die and other problems ensue. When silicon has been added to steel to restrain eddy current losses by raising of resistivity the hardness is raised also by such additions. Consequently steel processed by method 1 may be hard enough to stamp conveniently. Higher silicon contents tend to be used for larger machines where the width of teeth is greater and the effect of unrelieved cutting stress is less. Fully organic coatings are also more in demand for large machines. It would seem that larger machines favour stampings from method 1 feedstock.

Where fully soft steel has been considered to be too soft for convenient stamping it has been popular to apply a final light cold rolling pass in the range 4–12% reduction in thickness to give a hardness appropriate for stamping. This process is variously termed a 'temper pass' or an 'extension pass' (length increases as thickness is reduced, density does not alter significantly and conservation of metal applies). Such a final rolling offers the opportunity to use rolls with a precision degree of surface roughness so that the final product is made just rough enough by contact with such rolls to facilitate gas penetration between layers of a stack receiving a batch anneal (decarburising). Further, such a rough surface minimises inter-laminar sticking during anneal.

When a temper pass is applied a final anneal is mandatory as the few per cent reduction which raises hardness gives unacceptable magnetic properties (raised power loss and deteriorated permeability). It is found that the strain energy put into the crystal lattice of the metal by extension passing has, under anneal, the effect of 'explosively' driving grain growth so that quite large grains grow during this treatment.

Figure 4.12 shows the dramatic impact of a few per cent extension on final grain size. The 'critical value' if exceeded leads to a less beneficial result.

The post-temper-pass anneal then becomes:

(a) Stress relieving
(b) Decarburising (if wet gas used)
(c) Grain growth promoting.

The degree of extension passing giving an optimal grain response depends on many factors, some economically driven so that metallurgical judgement is required to give an optimal out-turn. For the limited reductions given in 'temper extensions' use of a rolling mill can be avoided by use of a 'roller leveller'. In this device, steel

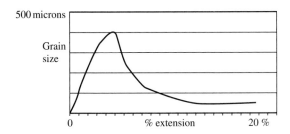

Figure 4.12   Plot of percentage extension versus grain size after anneal

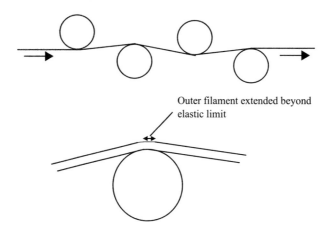

Outer filament extended beyond
elastic limit

Figure 4.13   Roller levelling

is passed successively over and under rolls of such a size that bending over each roll produces extension of the surface in excess of the elastic limit of the metal (e.g. some few cm diameter for 0.5 mm thick steel). This technique builds in strain energy which can drive critical grain growth. Figure 4.13 shows the technique. The rolls can be passive and the strip pulled through them, or actively driven either in concert or at selected different speeds. The location of a simple temper mill stand or roller leveller at the end of a strand anneal line is one of the options available to plant designers.

4.6   Hardness control

Steel hardness can also be raised by the addition of phosphorous in the range 0–0.08%. Too much phosphorus harms magnetic properties. So, hardness can be influenced by:

(a)   Percentage of silicon present
(b)   Degree of temper pass used, if any
(c)   Amount of phosphorous added.

4.6.1   *Perception of hardness*

From the above it may be thought that 'crisp' steel is a *sine qua non* of stamping. However, practice varies round the world. In Japan and the Far East the stamping of organic coated fully soft steel as a last operation has been a long-established practice and is well-understood in that quarter. In the USA and Europe, particularly the UK, the stamping of 'crisp' steel has been favoured and a wealth of expertise built up in this practice. Historically the development of cold rolling and strand annealing in the USA led to the use of critical grain growth as a route to enhanced magnetic properties. The fact that the need to stress-relief-anneal last could be combined with decarburisation is very convenient. The ability to decarburise during strand anneal gave an extra dimension to these practices. Finally the advent of vacuum degassing and ultra-low carbon steel completed the picture. If ultra-low carbon steel (as from steel-making) is wet gas annealed its properties can be impaired. The undesirable subsurface oxidation produced by wet gas annealing can be considerably restrained by the addition of small amounts of antimony (Sb) to steel, but antimony is a toxic element and so its use is not widespread.

Thus, decarburisation can be carried out:

(a)   At steel-making
(b)   During strand annealing
(c)   During final post-punching anneal
(d)   By a mixture of (a), (b), (c), designed to fit in with production facilities and economics.

Strain energy can be incorporated into a steel crystal by means other than cold rolling, such as roller levelling.

4.7   Box annealing

Strand annealing lines are expensive items and there has long been a practice of annealing steel in coil form using a batch furnace. Normally several coils form a box furnace charge (Figure 4.14).

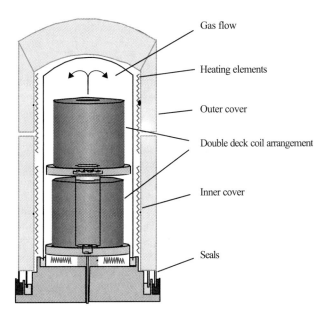

Figure 4.14   Cross-section of a box anneal furnace

4.7.1   Advantages of box annealing

The basic plant occupies less space and is less capital intensive. The tonnage through-put can be greater with the ability to use the box furnaces for other products. The deployment of larger amounts of specialist atmosphere gases is avoided.

4.7.2   Disadvantages

A batch process is slow. Time must be allowed for the least favoured portion of a coil to obtain the required temperature, and remain at it, for sufficient time. Excessive temperature can provoke lap-to-lap welding of coils which makes expensive scrap. The properties of annealed strip will not always be uniform. For instance the outer laps of a coil could be over-heat-treated while the middle remains under-annealed.

Decarburising is not a practicable feature (open-coil-annealing has long gone). Box annealing can be designed to give good results provided that a close study is made of the variability of properties that can arise and steps are taken to keep this variability within an acceptable range.

4.8   Furnace management and method A

Within the steel-producing industry the production and application of furnace atmospheres is a refined art. The 'raw materials' are:

(a)   Source of very pure dry nitrogen
(b)   Source of very pure dry hydrogen
(c)   Means of adding water vapour to gases to raise dewpoints and control them in the range 0–80 °C (dewpoint).

The technology to continuously measure gas dewpoint is vital and a range of commercial instruments are available. Industrial gases are usually prepared by elec-trolysis, air distillation or molecular sieving. In general the steel-producing industry operates with bulk sourcing of pure gases. Heat treatments carried out by laminators tend to operate by one of two routes:

(A)   Atmospheres derived from bulk supplies of industrial gas of a pure character
(B)   The use of atmosphere gas derived from the combustion of fuel gas.

In the case of (A) nitrogen can be obtained and stored as liquid nitrogen. Hydrogen can be bought in bottled form. In the USA more and more liquid hydrogen has become a 'commodity' and is a readily purchasable feedstock. The possibility exists of using a co-dissolved mix of liquid hydrogen/nitrogen which will deliver an evaporation mix of constant and convenient composition.

Liquid ammonia can be the basis of a furnace atmosphere supply since when catalytically cracked yields a mixture of three parts hydrogen to one of nitrogen by volume (NH_3).

Industrial cracking plants are readily available and may be fed from liquid ammo-nia vessels. Tanker delivery of liquid ammonia is industrially convenient. Bulk supplies of pure dry hydrogen can also be derived from cracking of the by-products of various petrochemical operations.

Pure gases produced as above can have very low dewpoints, e.g. −80 °C, and may be considered as 'dry'. Such gases are excellent for the heat treatment of ultra-low carbon steel where oxygen levels are kept as low as practicable.

4.8.1   Method B

This involves the combustion of fuel gas. Fuels such as natural gas, bottled propane, methane, etc. can be used. When burnt with air in carefully controlled reactions the combustion products contain a mixture of components, one of which is water vapour. Carbon monoxide and carbon dioxide will also be present.

Such a combustion source is, for historical reasons, referred to as 'exothermic gas' or 'exo-gas'. The control of the dewpoint of exo-gas can proceed by:

(a)   Drying it down to a very low dewpoint (freeze/condense out water). Then it may be used dry for anneals not required to be decarburising, e.g. for ultra-low carbon steel.

(b)   Dry down as in (a) then re-wet up to a specifically regulated dewpoint in a saturator.

(c)   Use direct from combustion with efforts made to control combustion gas/air mix ratios, etc. This is often the feed gas used for decarburising anneals and is the most difficult to control precisely.

4.9   Saturators

Figure 4.15 shows a saturator in diagrammatic form and Figure 4.16 shows laboratory scale and industrial scale saturators, and later sections give some notes on the management of these processes. Most usually the producer of laminations will use a continuous-batch method of lamination annealing.

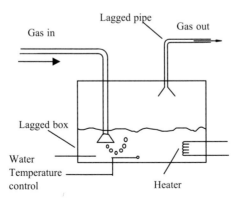

Figure 4.15   Outline diagram of a saturator

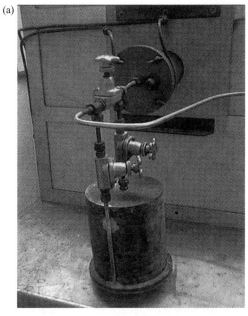

Figure 4.16   *a*   Laboratory saturator (insulation removed)
                b   Large industrial saturator

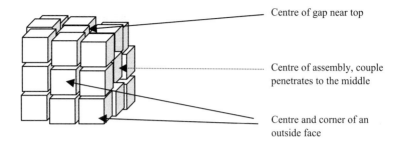

Centre of gap near top

Centre of assembly, couple penetrates to the middle

Centre and corner of an outside face

Figure 4.17 Assembly of lamination stacks. When loaded into a frame-tray in this way heat and gas has poorest access to items in the centre.

Stacks of laminations are loaded into trays (Figure 4.17) and these are passed mechanically through a furnace divided into sections. A preliminary heating drives off stamping lubricant, then the trays enter a zone with atmosphere control in which the charge is heated to annealing temperature, e.g. 790–800 °C. A soak time at temperature might be 1 hour or so. Trays are then cooled, still in a protective atmosphere, till at about 400 °C or less they are exposed to air and maybe to forced cooling. The transition from stage to stage is via doors/curtains to give adequate isolation between process stages.

Points of concern are:

(i)   The combined effects of heating and soaking should ensure that the least advantaged lamination in the centre of a charge receives an adequate time/temperature cycle. It is important that those most exposed are not overheated. The time at temperature may have to be longer if gas penetration is required to perform decarburisation.

(ii)   Exposure to air must not be too early as this promotes sticking between laminations and raised power losses.

(iii) Steel surfaces need to be slightly rough to facilitate gas penetration between laminations and to minimise sticking, but not be so rough as to greatly affect the stacking factor and permeability. An irregular surface provides flux paths inconsistent with best 'streamlined' flux flow, and this reduces permeability. The progress of laminations through a heat treatment cycle can be monitored by the use of trailing thermocouples whose junctions are placed in selected spots in a charge tray. However, the operation of inter-section doors tends to sever couples so that the use of a 'Squirrel' is required.

'Squirrel' is a trademark of Grant Instruments (Camb) Ltd., though it is sometimes used informally in a generic way. Similar devices called 'Datapaqs' are made by Datapaq Ltd in Cambridge. This is a data recorder contained in a heavily insulated box able to survive passage through a lamination heat treatment cycle. Such a device is put with a steel charge in a treatment tray and thermocouples from it disposed appropriately within the steel charge. When the device emerges at the end of a cycle it is cooled and opened and the data downloaded from the recorder. Pre-cooling of squirrels before insertion can be applied. Squirrels and Datapaqs are expensive and in the case of a furnace stoppage can well be destroyed by overheating.

The management of laminations calls for the following elements to all be present:

1. Understanding of the reasons for and design operation of the plant
2. Good equipment well maintained
3. Comprehensive measuring devices, kept well calibrated; e.g. thermocouples, gas analysers, Squirrel or Datapaq owned or rented, dewpoint meters
4. The keeping of comprehensive quality records which enable correlation between the final quality of steel and the operational parameters of the steel.
5. A programme of training and updating of information for operators/managers.

If the furnace is ignored until trouble becomes evident the cost of remedies can be high, involving lost production, extensive trial and error, etc. At the lowest level appropriate operating precautions allow good performance to be combined with good fuel economy.

A close grasp of the cycles which have to be achieved permits appropriate training to be instituted with consequent fuel economy and improved productivity. Particularly sticking is to be avoided or else crude methods of remedial treatment have to be applied, such as the application of mechanical shock to unstick laminations. Specifying a suitable coating on steel can allow good stamping to be combined with good sticking (lack of) behaviour. In reviewing the comparative merits of finally annealed steel and not finally annealed steel it may be noted that:

Finally annealed steel does not require a further heat treatment to develop its magnetic properties. Grains have already been grown during a special final heat treatment. But, if stamping is applied to FA steel this will degrade magnetic properties unless an extra stress relief anneal is applied. The larger the lamination the less the impact of unstress relieved cutting as the cut-affected zone forms a smaller percentage of the whole metal volume, i.e. wider teeth have more of the tooth remote from the cut edge. Decarburisation cannot be extensively performed in a grain growth anneal but will probably have been done at some point before the end of factory processing. However, final anneal heat treatment facilities are not required (capital or variable costs).

Not finally annealed steel requires a final anneal to grow grains and maybe to perform decarburisation, but this anneal will be after stamping, giving the best properties on small laminations. However the owning and operation of annealing plant is required. It may be noted that while the punching of NFA (not finally annealed) steel can be seen as easier (in USA and Europe) due to moderate 'crisp' hardness, FA (finally annealed) steel is likely to have a higher silicon content to keep eddy current losses low (valued on large machines). Thus it is automatically harder due to this added silicon and responds reasonably to punching in a fully annealed state. Further, makers of small machines where absolute losses are less vital may prefer very low silicon material so that the highest $\hat{B}$ value gives higher torque in a machine. In every case a designer needs to appraise the benefits to his machine which arise from the use of FA or NFA feedstock. Best results flow from incorporating all relevant factors and costs into initial design work.

4.10　Blueing

When hot steel is exposed to an atmosphere containing oxygen, free or as H_2O, a blue (or other colour depending on conditions) film forms on the surface. This blue film can be the basis of a rudimentary coating and is often carried out as one of the operations in an annealing cycle. Here steam may be injected into the system as a separate operation and the familiar blue coloration follows. Steam blueing is cheap and easy to apply, but does not have the same level of controllability and efficacy as a purpose-designed coating.

A brief review of annealing will remind the reader that the motives for annealing include:

Homogenisation of micro-chemistry
Production of phase changes
Control of grain size
Influencing of grain orientation
Removal of stress
Flattening
Decarburisation
Control of hardness.

Anneals can be:

Strand continuous
Box – coils
Box – laminations
Continuous/tray
Mesh belt.

Atmospheres can be:

Neutral dry
Decarburising wet
Mixtures of these.

4.11   Summary

The main processes described in this chapter may be set out succinctly as follows.

Motor lamination steel (steel not finally annealed at delivery)
Input: Hot rolled coil approx 2.0 mm thick, no added silicon.

Process	Purpose of process
Shot blast, possibly add oil and side trim	To remove hot mill scale and lubricate prior to cold rolling. Provide perfect edge for cold rolling.
Cold roll to intermediate gauge	To reduce the material to the correct intermediate thickness and provide good 'shape'.
Intermediate strand anneal in decarburisisng line at about 800 °C, 50–60 mpm decarburising atmosphere.	To recrystallise the metal and decarburise if required.
Cold roll to give 3–8% length extension, using roughed rolls, final thickness usually 0.65 mm.	To provide a critical amount of surface strain and to impart a specific surface roughness and apply a corrosion inhibitor.
Side trim, slit, coil up.	Prepare final form to customer requirement.
Final anneal by customer after stamping.	To develop correct grain size and optimum magnetic properties as well as removing internal stress.

This type of steel has no formal coating and the roughened surface prevents sticking of laminations during the annealing of a stack and allows gas penetration if decarburisation is needed in the final anneal.

Medium motor steels

This process (above) can be applied to steel with added silicon up to some 2.0%. The added silicon raises resistivity which reduces power loss. High field permeability will be reduced but this is acceptable when low iron loss is essential for larger machines.

Polycor (not fully annealed – low silicon rolled coating)
This type of steel has been specially formulated for the production of energy-efficient versions of traditional induction motors.
Input: Hot rolled coil approx. 2.0 mm thick with about 0.25% silicon and ultra-low carbon and ultra-low sulphur.

Process	Purpose of process
Strand anneal the hot-rolled coil, temperature up to about 1000 °C and pickle, oil and side trim it.	To anneal the material and develop the required hot-rolled grain structure. To remove hot mill scale and to apply a lubricating oil film.
Cold roll to intermediate gauge.	To reduce to correct intermediate thickness and get flat 'shape'.
Intermediate anneal in a strand anneal at about 800 °C and apply an insulating coating of a type able to survive further cold rolling.	To recrystallise the material and apply and cure a mixed organic/inorganic coating.
Cold roll to give about 6% extension of length. Final thickness usually 0.5 mm. Side trim, slit and coil up.	To provide appropriate strain energy to drive critical grain growth. Prepare final form to customer requirement.
Final anneal by customer after stamping.	To develop correct grain size and optimum magnetic properties as well as removing internal stress.

Large machine lamination steel (fully annealed)
Input: Hot rolled coil approx. 2.0 mm thick with added silicon ranging up to 3%.

Process	Purpose of process
Pickle, shot blast, side trim and oil.	To remove shot mill scale and prepare material for cold rolling.
Cold roll. Final gauge may be 0.50, 0.35 mm.	To reduce the steel to correct final thickness and flat shape.
Strand anneal at about 900–1050 °C and apply coating.	To recrystallise the material and produce critical grain growth for optimisation of magnetic properties. The coating provides inter-laminar insulation and normally organic, though mixed coatings may be used.
Side trim, slit, coil up.	To prepare final form for delivery.

Chapter 5

Coatings and insulation

It has become clear from earlier discussion that the lamination of cores is needed to restrain eddy currents and to allow rapid flux penetration into the bulk of a core. Such lamination is only efficacious if adequate insulation is maintained between layers. The application of a suitable coating can ensure that insulation between sheets is adequate. There are other motivations for having a coating present. These are:

(a) Appropriate coatings promote a better punching performance (low burr, long tool life). The precise mechanism of punch-tool 'lubrication' by coatings is complex, but the result is clear – organic species greatly facilitate stamping.

(b) Restraining sticking between laminations (avoids metal/metal welds). When sheets are pressed together at high temperature ($500+ °C$) actual interpenetration of crystal atomic structuring can occur and lead to micro-welds which make laminations hard to separate and facilitate the passage of eddy currents.

(c) Allowing a specific range of surface sliding friction to be attained – high or low. Low interlaminar friction is useful for the slick handling of stator stacks and facilitation of their alignment during core assembly. However, when self-supporting stacks are squeezed under pressure the mechanical stability of the pile is better when moderate surface friction is available, until full mechanical restraint can be applied.

(d) Forming the basis of a 'consolidate-into-a-block' process when heated and squeezed in a core assembly process. It is convenient to use the insulating coating as the source of interlaminar 'gluing' such that the application of heat and pressure to a core stack converts it into a rigid self-supporting block.

There are however hazards of having a coating involving:

(a) Generation of gas and the production of blowholes in lamination stacks which are bead-welded (see Figure 5.1). Bead-welding is a popular method of consolidation of lamination stacks, and while the gas evolved from the decomposition of the organic component of a coating can give blowholes the physical security of the stack may be little affected with only aesthetics suffering.

(a)

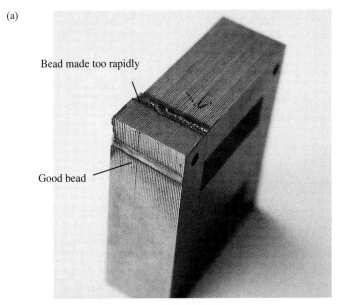

Bead made too rapidly

Good bead

(b)

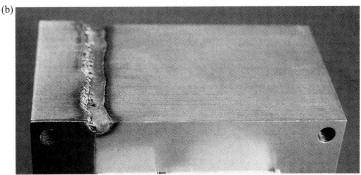

Figure 5.1   a   Good and bad weld beads
            b   Weld bead with blowholes

(b)   The penetration of gas into stacks of laminations and onto the steel surface, if decarburisation is needed, may be impeded by some coatings. In general, however, coatings become adequately porous at high temperature.

(c)   Delivery of carbon to the lamination under inappropriate heat treatment. The decomposition of organic coatings can be a source of carbon, but serious effects are rarely seen.

When coatings are used they have to withstand the operating and processing conditions of the final core. This can involve:

(a)   Exposure to high temperatures 120–800 °C for varying times. Some motors must work briefly up to 500 °C during emergency operation.

Figure 5.2 *Points of highest electric stress A and B. If A is shorted then stress doubles at B*

(b) Exposure to attack by solvents (e.g. hermetically sealed refrigerator motors). The progressive evolution of refrigerants (anti-greenhouse) means that the effect of new ones on coatings has to be frequently assessed.
(c) Exposure to corrosive atmospheres, e.g. wet, salt spray. Environments associated with machines in motor cars, chemical plant, etc. pose special threats to steel coatings.
(d) Exposure to vibration/abrasion. Whenever possible relative motion between adjacent lamination plates should be avoided to reduce the risk of scrape-off of coatings.

The width of laminations has to be considered. Examination of Figure 5.2 shows that the wider the lamination the higher the electric stress on the coating. Until laminations reach some 100–200 mm wide the effect of a coating on interlaminar currents is small. A study has been made of interlaminar insulation for small motors and a brief review of this project is given in Chapter 10.

A review of coating types can be made as follows. The three main types are:

- Fully organic coatings
- Mixed organic/inorganic coatings
- Fully inorganic coatings.

Of course, no-coating-at-all represents an informal further group in which natural surface oxides provide some sheet–sheet isolation. Table 5.1 outlines the properties and applicability of these main groups.

5.1 Nature and methods of application

5.1.1 Fully organic coatings

These are of the resin-emulsion type and are applied by roller coating using water as the carrier component of the emulsion. This type of coating can be applied where wide components (high interlaminar emf) are likely to be encountered. Thicker-than-usual coatings may act to 'bury' shearing burrs and prevent short circuits (Figure 5.3). In large machines supplementary varnish may be added last to double-protect against unwanted currents. The nature of organic coating is such that it can act to lubricate a die and punch combination so that tool lifetime is greatly extended.

Table 5.1   Properies of SURALAC® coatings for finally annealed non-oriented electrical steels

Designation	SURALAC 1000	SURALAC 3000	SURALAC 5000	SURALAC 7000
Type	Organic	Organic	Semi-organic	Inorganic
Description	Organic phenolic resin	Organic synthetic resin with inorganic fillers	Organic resin with phosphates and sulphates	Inorganic phosphate based coating with inorganic fillers and some organic resin
Old Surahammar designation	C-3	C-6	S-3	C-4/C-5
AISI type (ASTM A 677)	C-3	C-6		C-4/C-5
Thickness range, per side	0.5–7 µm 20–280 µinch	3–7 µm 120–280 µinch	0.5–2 µm 20–80 µinch	0.5–5 µm 20–200 µinch
Temperature capability in air (continuous)	180 °C 355 °F	180 °C 355 °F	200 °C 390 °F	230 °C 445 °F
Temperature capability in inert gas intermittent	450 °C 840 °F	500 °C 930 °F	500 °C 930 °F	850 °C 1560 °F
Withstands:				
Stress relief annealing	–	–	–	YES
Burn-out repair	–	YES	–	YES
Aluminium casting	YES	YES	YES	YES

SURALAC®	1007	1025	1060	3040	3060	5007	5012	7007	7020	7040
Typical thickness (µm) per side	0.7	2.5	6	4	6	0.7	1.2	0.7	2	4
(µinch)	30	100	240	160	240	30	45	30	80	160
Typical welding	good	spec	spec	spec	spec	exc	exc	exc	good	mod
Typical punching	exc	exc	good	good	mod	good	exc	good	good	mod
Surface insulation resistance (Franklin ASTM A 717)										
Typical value, ohm cm² per lamination	5	50	>200	100	>200	5	20	5	50	100
Typical value, ampères per side	0.55	0.11	<0.03	0.06	<0.03	0.55	0.25	0.55	0.11	0.06

Reproduced with permission: European Electrical Steels.

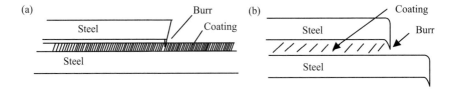

Figure 5.3   *a*   Burr 'accommodated' in coating
            b   Benefit of supplementary varnish

5.1.2   Mixed coatings

These involve both organic and inorganic components. Usually a mixture of polymeric resins and inorganic material is used.

   These coatings are applied by a roller coating method in a water base and the whole is dried and cured in a continuous strip furnace. This can all take place at the end of a strand-anneal line or be part of a free-standing coating set-up.

5.1.3   Fully inorganic coatings

These are mostly used for transformer steels where stamping as such is rare. Some motors requiring to operate at high temperature, e.g. 400 °C either routinely or in an emergency, may specify this type of coating. Coatings are applied by roller coating, then cured at about 800 °C in a continuous strip line. Most are based on phosphates with other additives. Stamping of fully inorganic coatings is carried out, but it is accepted that even with tungsten carbide tools, tool life is much lower than for other coatings.

5.2   New developments – print-on coatings

Some years ago I considered that energy and space could be saved by using print techniques to apply coatings. 'Inks' require no free solvent carrier so that drying of a water-based coating is avoided, and the environmental risk from using organic solvents is absent. By using a print-on technique the energy costs of drying and curing can thus be reduced. To coat electrical steel a coating need only be of one colour and the complexity of readable printing is not involved. After many trials a successful print-on coating has been devised and is now a fully useable product.

   A certain degree of mystique surrounds the formulation and application of coatings as commercial confidentialities are involved. Users of steel are often reluctant to accept coatings which may look different from those which they are used to receiving, even if the technical properties are demonstrably superior. When coatings have become obsolete and been replaced by better (and cheaper) varieties the change has often been met with deep conservatism on the part of the user. This situation is best addressed by providing as much quantitative information about functional properties as possible rather than placing reliance on aesthetic appearance.

Many coatings can be applied via roll-coating systems which are fed by either water-based or organic solvent-based carriers. Water-based coatings require a relatively large amount of heat and time to 'dry' and cure during progress through a strand curing oven which may be either free standing or concatenate with the end of a strand-anneal line.

The capital cost of such an installation is high and its energy demands and space requirements are considerable, e.g. 30–100 metres long for coat and cure. Organic solvent-based coatings have the disadvantage that the use of solvents presents hazards of fire risk and environmental pollution if products of curing escape. Organic solvent-based coating systems are now actively discouraged for new installations.

European Electrical Steels have devised and pioneered a coating technique which overcomes many of the usual problems. This technique is called print-on coating, referred to above. It is well known that the print industry has long-standing skill in printing onto many substrates including metal. These can range from newsprint to biscuit tins. By drawing on print techniques that offer an even thin coating in one 'colour' alone an electrical steel insulant is produced. Curing can be achieved very rapidly by the application of ultraviolet (UV) light immediately after printing.

Development of this technology has enabled print-on plus cure to be compressed into a few metres of line length. If the curing section is formed into a 'U' loop then the on-the-ground length falls even further. This system is thus compact and requires less energy to operate than regular devices.

UV curing relies upon the principle that the ink contains suitable monomers which cross-link rapidly into polymers when UV light is applied. For thin layers, as are appropriate for electrical steel insulants, self-absorption of the UV is insufficient to inhibit full curing through the thickness of the printed coating by the outer layers of coating shielding that part nearer to the steel surface. Up to 15 microns can be penetrated satisfactorily.

Although electron beam curing can perform as well as UV light it is not a convenient system for production line use at atmospheric pressure as the range of energetic electrons in air is small. High-pressure mercury lamps provide an appropriate source of ultraviolet light. Some ionisation of air is produced by UV light and ozone is formed. This has to be removed from the working area by an appropriate ventilation system since ozone, O_3, is a reactive gas which presents hazards.

The actual coating system used involves the take-up of a relatively viscous resin by Anilox rolls whose surface is formed so that very many tiny cells or pockets harbour ink (Figure 5.4). Cell dimensions are of the order of 60 microns across and 15 microns deep. A 'doctor' blade removes surplus ink leaving only the cell fillings. This is transferred via a rubber applicator roll to the moving strip after which curing is applied. The amount of 'ink' transferred is controlled by the Anilox cell pattern and can be precisely controlled. Exchange of rolls is the method used to alter coating thickness. Coatings in the range 0.5–6.0 microns are currently produced.

Figure 5.5 shows the effect of printed coating thickness, for a typical coating, on insulation values (measured by a Franklin tester), and relates to the aggregate of top and bottom surfaces of the steel.

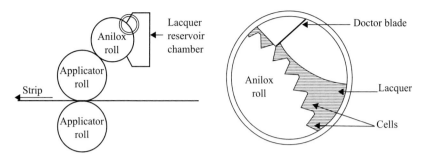

Figure 5.4   Anilox roll system

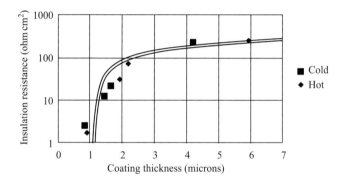

Figure 5.5   Insulation resistance data showing the effect of thickness

One of the aims of coating development is to produce coatings which function well both before and after annealing treatments without recourse to chromium compounds (which are toxic for certain valency states of chromium). Results of print-on trials are very promising in this regard, along with resistance to the usual solvents. Coatings of good weldability are selectable from the range of print-on coating materials. The punchability of printed-on coatings is much superior to that of bare steel. Figure 5.6 illustrates this point: it refers to punching with an alloy steel punch and die, not carbide tools.

The prospects for print-on coatings are very good as they offer coating application in a compact factory space with low energy demands and are environmentally friendly both in terms of no solvent release and in the absence of chromium.

Electrical and punching performances are good, as shown (Figures 5.5 and 5.6).

Table 5.1 outlines some of the coatings produced by European Electrical Steels (now Cogent Power Ltd.).

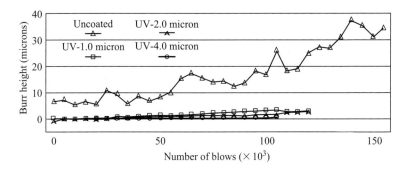

Figure 5.6   Burr height versus number of blows

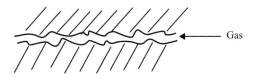

Figure 5.7   Rough surfaces allow gas penetration

5.3   Blue oxide coatings

The production of blue oxide coatings has already been considered in Chapter 4 where such a coating can arise as part of the heat treatment and decarburising process. Although blue oxide offers very low insulation resistance, this is often considered to be sufficient in the case of small laminations where the interlaminar emf is low. Blued surfaces are a little more corrosion resistant than bare steel, and when laminations are blued as a last process the cut edges are also protected.

The consistency and extent of bluing can be well controlled if a dedicated chamber receiving steam injection is used at a later stage of the decarburising process. A slightly rough steel surface, e.g. 0.6 microns (25 micro-inches), is needed to allow bluing gas to penetrate to all surfaces (Figure 5.7). This will have been needed for decarburising anneal, but in the case of ultralow-carbon steel, penetration of decarburising gas is not needed and bluing is not helpful as it can contribute to subsurface oxidation.

In every case decisions have to be made as to the spread of material types to be put through a process and the effects it will have on each.

Ultralow-carbon steel, while more expensive, is amenable to simplified non-decarburising heat treatment and it may be useful to meet all needs from one feedstock.

5.4   No coating at all

It has been found that for small machine laminations, e.g. up to some 20 cm diameter, the interlaminar emf is so small that ohmic currents of significant intensity will not flow unless actual crystal interpenetration has occurred. This could be at a sticking weld or beneath a crushed burr.

The use of a coating is not needful if burrs are restrained and micro-weld sticking is avoided. Coatings are, however, useful to provide die lubrication, double-ensure against sticking, and to provide slip-ability of stacks of lamination which may need to be manipulated before anneal or core assembly.

5.5   Rolled coatings

Usually coatings are applied to electrical steels as a separate operation often on a separate production unit. In the case of semi-finished steels (where a coating may be desired to avoid sticking of lamination stacks) the last production operation before such coating is cold rolling. Normally such cold rolling employs water-based mill lubricants and before coating can be applied the steel has to be cleaned and dried. This is an expensive operation, as is the extra handling of material onto a coating line.

I considered that it should be possible to devise a coating and rolling regime that would allow the coating to be applied as the last in-line operation of a strand annealing or annealing/decarburising line. Here the strip is dry and perfect for coating as it emerges from the strand-anneal furnace.

Passing strip coated in this way onwards into a cold reduction mill to receive the final 6–8% extension pass appeared impossible as coatings may be expected to crumble under rolling and become ineffective as well as releasing detritus which would block up the mill cooling system. However, persistence with experimentation led to a practice in which a coating could be designed to survive rolling so that the final product was coated, then 6–8% extended. It then also carried a light covering of corrosion inhibiting rolling fluid.

It has become common practice to include corrosion inhibitors (anti-rust) in final-stage rolling fluid to provide some corrosion inhibition for steel coils during transport and warehouse storage.

Table 5.2   Summary of AISI coatings (USA scheme): Surface insulation for electrical steels

Type code	
C-0	Natural oxide arising without use of a specific coating procedure. It may be enhanced by oxidation occurring during a stress-relief anneal.
C-2	Inorganic glass insulation for grain-oriented steel. Annealable.
C-3	Enamel/varnish, not annealable.
C-4	Phosphate or similar, able to withstand stress-relief annealing.
C-5	Supplementary coating applied over C-2 type. Can be neutral-atmosphere annealed.

After lengthy development this practice is now well-established and patent-protected. A mixed organic–inorganic coating is used for rolled coatings so that in subsequent annealing treatment the inorganic component remains to provide insulation while the organic component promotes good tool life during stamping.

5.6   Formal classification of coatings

The most widely used classification system is the AISI (American Iron and Steel Institute) description. This is summarised in Table 5.2 Composite coatings are created when grain-oriented steel has an extra phosphate layer applied on top of the magnesium silicate glass formed as part of the grain orientation process. Composite phosphate coatings are not, however, kind to stamping tools. This has a high ohmic resistance: a standard performance expected of this type of coating is better than $10 \, \Omega \, cm^2$.

5.7   Coatings and the small motor experience

A discussion of the impact of insulation coatings on small motors is given in Chapter 10 and [5.1].

5.8   Punchability

The presence or absence of a coating and its nature has a profound impact on punchability. This will be considered under burr measurement in Chapter 10.

5.9   Corrosion resistance

Freshly rolled and favourably heat treated steel corrodes very rapidly in a moist atmosphere. Generally steel will be finished so that a corrosion inhibitor becomes applied during the last production process, or the storage and packing is organised so that ingress of moisture is prevented and an environment of vapour phase corrosion inhibitor is present. However, the working environment of core laminations may present a special 'corrosion' problem. Hermetically sealed refrigerator motors must work in continuous contact with refrigerants. The emergence of new refrigerants in response to the Montreal Protocol requires that solvent response must be continually updated.

Widespread testing is carried out to verify appropriate resistance of steel coatings to a wide range of solvents. Also stamping presses may use water-based lubricants whose effect on coatings may have to be noted.

5.10   Heat resistance

In some applications motors are required to continue in operation at least for a time under very adverse conditions of environment or sustained overload, for example in military of fire safety equipment. Coatings may be specified to withstand 800 °C for long periods even though usual duty is at room temperature. In such cases punchability has to give way to special features of specification.

From time to time machines fail and require repair. This usually takes the form of 'burn out' in an autoclave.

Windings are often secured in place by epoxy resin and this has to be decomposed to permit removal of the old windings. Unfortunately not all operators in this field are as careful as they could be. To minimise damage to interlaminar insulation it is important not to overheat the core, particularly if oxygen is not properly excluded.

It is known that oxygen permeating a core at temperatures above 400 °C can promote interlaminar adhesions which allow electron flow. None the less the presence of a good-quality coating restrains most of these ill-effects.

Figure 5.8 compares the effect of 'burn out' temperatures on losses in rebuilt machines. Ref. [5.2] discusses this. Clearly an appropriate coating is valuable in the repair situation.

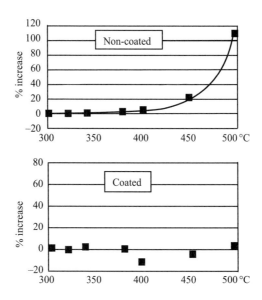

Figure 5.8   Effect of 'burn out' temperature on core loss

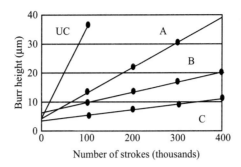

Figure 5.9   Punching characteristics of uncoated steel and three particular coatings A, B and C

5.11   Stacking factor

Thick coatings devised to provide burr hiding on larger machines, especially if accompanied by additional coating applied after stamping, reduce the effective space occupancy (stacking factor) of core steel. A lowered stacking factor reduces the effective high-field permeability of the core space and eventually raises copper losses because more magnetising current is needed to realise the same flux output.

5.12   Weldability

It may be stated that a weld bead that shows blowholes (Figure 5.1b) is just as secure a fixing mechanism as a perfect bead, but such are the aesthetic demands of end users that bubbled welds may be rejected. Features that improve weldability (e.g. low organic content) tend to impair punchability as indicated by Figure 5.9.

5.13   Comments on coatings

SURALAC 1000 is an organic coating (C-3 type) which has excellent punching characteristics and can be supplied with a range of insulation resistance values depending on coating thickness.

SURALAC 3000 is an organic coating with fillers (C-6 type) which has good punching characteristics, very good insulation resistance which is dependent on coating thickness, good burn-out and compressibility properties. It can also be applied as a thick coating.

SURALAC 5000 is a semi-organic coating with very good punchability and very good welding response.

SURALAC 7000 is basically an inorganic coating with organic resins and inorganic fillers (C-5 type) which has excellent welding and heat resistance properties with good punchability. Insulation resistance is again a function of coating thickness.

Selecting a coating for a particular application, e.g. segments for a large turbo-generator, laminations for a small induction motor, etc. can be a complex procedure, as a worthwhile coating specification must also take the manufacturing process into account. As mentioned above, the final decision may have to be a compromise based on a comprehensive study of coating properties. The decision-making task is not made any easier by a shortage of factual information from the user, e.g. interlaminar emf/currents in machines. On occasions, a scientifically valid choice can also be hindered by a conservative attitude which will dictate a selection based on traditional practice. On the other hand, there is always some hesitation in revising a long-standing specification which 'works'.

References

5.1. BECKLEY, P., LAYLAND, N. J., HOPPER, E. and POWER, D.: 'Impact of surface coating insulation on small motor performance', *Proc. IEE Elec. Pow. App.*, 1998, **145** (5), pp. 409–13.
5.2. COOMBS, A., LINDENMO, M., SNELL, D. and POWER, D.: 'A review of the types, properties, advantages and latest developments in insulating coatings on non-oriented electrical steels', Proceedings of IIT Conference on *Magnetic materials*, Chicago, May 1998.

Chapter 6

Range of materials

Materials are made, specified and used against a pattern of properties and applications.

6.1 Properties

(a) Composition. Added silicon can range from near zero up to 3+ %. In general higher silicon which restrains eddy currents reduces loss, is harder and costs more to make. Resistivity and as-annealed hardness varies with the percentage of silicon. Added silicon reduces B_{sat} and impairs high-field permeability. The absence of carbon, sulphur and titanium is desirable; desulphurisation is possible but expensive.

(b) Thickness. The thinner the steel the more effectively eddy currents are restrained and the lower the core losses. However a lamination pile of a given height contains less metal if made of 0.35 mm steel rather than of 0.65 mm steel due to the effect of the extra surfaces on the stacking factor. The effect of this is to reduce the effective B_{sat}. Thinner steel costs more to produce and requires more press-blows to make lamination stacks of a given size.

(c) Coated or not. Coatings are expensive to apply. Needless coating reduces the space factor and may cause weldability problems.

(d) State of heat treatment. Material can be finally annealed or semi-processed and awaiting a final anneal after stamping.

(e) Hardness. Material can be soft due to being finally annealed and having very little alloy content, e.g. VPN_{10} 90, or, progressively harder, as:

- Silicon is added up to 3%
- Phosphorous is added up to 0.08%
- Temper rolling is applied.

Hardness up to VPN_{10} 180 is acceptable. Note: the subscript 10 refers to a test done with a 10 kg load.

Table 6.1   Selection of electrical steel grades produced by European Electrical Steels – typical properties

Grade identity	Thickness (mm)	Typical specific total loss* (W/kg)	Typical specific apparent power* (VA/kg)	Nominal silicon content (%)	Resistivity (Ω m $\times 10^8$)	Stacking factor	$\hat{B}^\dagger$ (T)	Typical applications
Grain oriented: magnetic properties measured in direction of rolling only								
Unisil H M103-27P	0.30	(0.98)	1.40 (1.7, 50)	2.9	45	96.5	(1.93)	High efficiency power transformer
M111–30P	0.30	(1.12)	1.55 (1.7, 50)	2.9	45	96.5	(1.93)	
Unisil M120–23S	0.23	0.73	1.00	3.1	48	96	(1.85)	
M130–27S	0.27	0.79	1.10	3.1	48	96	(1.85)	
M140–30S	0.30	0.85	1.13	3.1	48	96.5	(1.85)	
M150–35S	0.35	0.98	1.24	3.1	48	97	(1.84)	
Non-oriented: magnetic properties measured on a sample comprising equal numbers of strips taken at 0° and at 90° to the direction of rolling								
Fully processed electrical steels								
M300–35A	0.35	2.62	23	2.9	50	98	1.65	Large rotating machines
M400–50A	0.50	3.60	19	2.4	44	98	1.69	Small transformers/chokes
M800–65A	0.65	6.50	14	1.3	29	98	1.73	Motors and fractional horse power (FHP) motors

Non-oriented: grades supplied in the 'semi-processed' condition and which require a decarburising anneal after cutting/punching to attain full magnetic properties

Newcor M800–65D	0.65	6.00	8.6	Nil	17	97	1.74	Motors and FHP motors
Newcor M1000–65D	0.65	7.10	9.6	Nil	14	97	1.76	
Polycor M420–50D	0.50	3.9	6.5	Nil	22	97	1.74	
Tensile grades								
Tensiloy 250	1.60			Nil			1.60	Pole pieces large rotating machines

* At $\hat{B} = 1.5\,T$, 50 Hz. Values in parentheses are for $\hat{B} = 1.7\,T$ and 50 Hz.
† At $\hat{H} = 5000\,A/m$. Values in parentheses are for $\hat{H} = 1000\,A/m$.

6.2   Application

Small machines generally demand a low-price steel. A steel at 0.65 mm unalloyed, awaiting final anneal with no formal coating could be typical. Larger machines benefit from a higher percentage of silicon and a definite coating. Very large machines use 3+ percent silicon and a good coating, perhaps followed by a further apply-last varnish when interlaminar emfs will be large.

6.3   Working conditions

The force between rotor and stator depends on B^2, so that for best power output per volume, higher B values tend to be used. If operating cyclic B_{max} values are too high then losses rise rapidly and permeability falls off, so that a compromise B_{max} value will be used. Careful attention should be given to the B_{max} at which purchase and sale are agreed. A steel having a lower loss at $B_{max} = 1.5\,T$ (a typical specification induction) may perform worse at $B_{max} = 1.8\,T$ if it has a higher percentage of silicon than an unalloyed grade would do. High silicon grades look best under test at medium inductions, but perform less well at higher inductions. Particularly in Germany, specifications at $B_{max} = 1.0\,T$ were common as a carry-over from long-past hot-rolled high-silicon steels and were confusingly inappropriate when applied to steel aimed at 1.5–1.8 T B_{max} operation.

It is possible to select a material of high quality (low power loss) and apply brutal fabricational methods to it, leading none the less, in spite of degradation, to a passable end-device. Perhaps a more careful (but costlier?) application route may have made the use of a more basic grade of steel acceptable. While more detailed grade-by-grade graphs of properties are set out in Chapter 15, Table 6.1 gives an overview of a set of materials spanning the property range. Included is a high-tensile grade which is used in those parts of large machines where mechanical strength is important but reasonable regard must be given to magnetic permeability since flux must be conducted.

6.4   The question of permeability

Clearly low power loss combined with good permeability is desirable, and high permeabilities are often specifically called for. However, efforts on the part of steel-makers to produce specially high permeability steels is sometimes met with indifference from machine makers. The view has been put forward that if the machine has, as it must, a stator–rotor air gap, then the reluctance of that gap will dominate the magnetic circuit and specially good steel permeabilities are of limited use (especially if attracting an extra cost). However, a contrary view is that if high tooth flux can be attained with a lower mmf then copper can be saved and space better used. Each designer must form his own view of this, but my opinion is that improved permeability can be usefully exploited and is worth a premium.

6.4.1 Expressions of permeability

In Europe permeability is usually expressed as the B-value attained when a specified H-value is applied, e.g. B_{H50}. If the application of 50 A/m gives 1.5 T, then B_{H50} or B_{50} is 1.5 T.

The USA practice is to quote the B/H ratio at, e.g. 1.5 T, and, using old units, if a B of 15,000 gauss is produced when 5 oersted is applied then $B/H = 3000$. This treatment involves a fixed B-value and tends to form a more useful shorthand.

Chapter 7

The effects of punching and core building

Ideally a punching process should take in steel strip and deliver perfect punchings of the same magnetic properties as the feedstock used. In practice a range of defects can arise during punching operations.

These are listed below.

7.1 Degeneration of magnetic properties due to shearing stress and crystal deformation

To cut sheet and form a shaped lamination, metal must be displaced and the crystal lattice disrupted. This produces regions of cold work, locked-in stress and dislocations. These sites pin domain walls, impair permeability and increase power loss. Typically the region of disruption spreads several mm from a sheared edge. If sharp tools are used with appropriate die clearances magnetic damage is minimised. The narrower a stamped item is, e.g. a stator tooth, the greater the percentage of its volume will be subject to punching disruption, so the effects have a bigger impact on small stampings.

Much of the magnetic damage can be recovered if appropriate stress relief annealing is applied after stamping, e.g. a few minutes at 800 °C in a neutral atmosphere. Stress relief annealing can be combined with a decarburising treatment if this is needed. Not all cutting damage can be recovered as the sequence of cutting and annealing can lead to a final grain pattern in the affected region that is less favourable than the original, if this was a large grain size carefully produced. Figure 7.1 shows the effect of cutting and annealing.

7.2 Production of burrs

Inevitably the process of cutting will displace metal so that a burr is formed on the cut edge (see Figure 7.2). The displaced metal forms a sharp edge which may well be

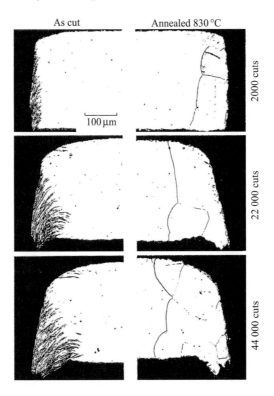

Figure 7.1   Sections through a cut edge in grain-oriented 30 mm steel as cut, and after stress relief annealing

forced into electrical contact with a sheet below it in a lamination stack. Much effort is expended to restrain burr and guard against its effects, but inevitably economic compromises must be made. In general sharp newly fettled tools give minimum burr when operated with appropriate punch-die clearances. There should be a match between the mechanical properties of steel and the punching tool design. If soft steel is punched with inappropriate tool spacings metal is 'smeared' so as to drag down into a longer burr.

The metal parameters relevant to good punching are:

(a) Hardness – VPN
(b) Ductility – percentage extension before fracture
(c) Ultimate tensile strength – UTS
(d) Yield strength, that is the tensile force required to produce a (e.g. 0.1 per cent) permanent deformation in the metal
(e) Yield/UTS ratio
(f) Presence or absence of die lubricants
(g) Presence or absence of coatings on the steel, which are favourable (organic) or unfavourable (inorganic) to burr minimisation.

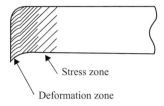

Stress zones at cut edge (section)

Figure 7.2   Burr formation

The whole subject of lamination stamping embodies a large volume of technical know-how, much of it commercially sensitive and outside the scope of this book. For instance tungsten carbide tools have a very long life but are very expensive to make. Their tool steel equivalents are cheaper, wear faster and are more suited to short runs of bespoke laminations.

The speed of press operation relates directly to the economics of lamination production. A thinner steel may have a better magnetic performance but require more laminations (more press blows) for a given stack height.

Best magnetic properties may not align with easiest punching, so that soft steel may be unwelcome to stampers. Punching at a stage before final softness is reached (e.g. as-temper-rolled) may be a favourable option. The addition of hardening elements, e.g. phosphorus, may help stamping, but not magnetic properties if overdone. The dynamics of punching involve the yield strength, UTS and yield/UTS ratio. Operational decisions relating to these factors lie within the area of competence of stamping managers. In essence the steel producer can offer an optimal magnetic product but should liaise with stamping experts to ensure that the best compromises on material properties be found for both steel manufacture and stamping. The steel producer must make products reasonably handleable by most stampers. Special stamping needs of small volume are not readily addressable.

Figure 7.3 shows the progress of increasing burr formation as stamping proceeds; clearly the nature of the coating on the steel has a great effect. These graphs are for a steel having a fully inorganic coating and using steel punching tools. When mixed or fully organic coatings are used burr development is delayed. Tungsten carbide tools give even longer tool life but at increased first cost. Factors such as press speed, etc. affect results so that beyond general trends producers of laminations must rely on their own experience of burr building in different situations.

7.3   Shape – stability

The punching of laminations is metallurgically a very disruptive process and can be upset by the following factors.

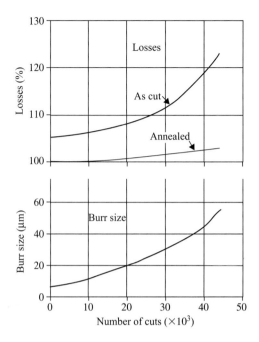

Figure 7.3   Relative losses and burr size in grain-oriented 0.30 mm steel as cut, and after stress relief anneal, versus number of cuts. Steel tools on 3 cm wide strip

7.3.1   Unbalanced stress during punching

Here the infeed of stress-free flat strip can nonetheless lead to laminations in which a round hole emerges as an eliptical hole in the finished lamination, stator teeth are deflected out of the plane of the lamination or the overall lamination is curved or dished. These effects arise from a mismatch between the strategy of punching used and the material properties. Stamping practice contains a lot of science and quite a lot of 'art'. Steel parameters affecting fidelity of punching, i.e. output of flat strip with round holes being truly round, etc. are:

(1)   Hardness of strip expressed as a VPN_{10} number both absolute for the composition of the steel and the extra hardness due to purposely unrelieved rolling stress
(2)   Presence of anisotropic internal stress
(3)   Grain size
(4)   Surface condition, oiled, coated or not, and what sort of coating
(5)   Absolute thickness of steel
(6)   Ultimate tensile strength
(7)   Proof stress
(8)   Proof stress/UTS ratio.

Punching parameters affecting punching fidelity are:

(1) Die clearance
(2) Speed of punching
(3) Degree of tool wear
(4) Applied lubrication
(5) Restraint of strip during punching.

Other factors include:

(1) Type of final anneal, if used
(2) Rate of temperature rise and cooling rate
(3) Temperature gradient in stacks.

No completely comprehensive manual of punching practice covers all situations and much depends on the experience of punching houses. Steel manufacturers try to provide parameter ranges requested by customers, but requests vary and a middle-of-the-road product has to be produced. Were it not for the relevance of commercial confidentiality and its impact on competitiveness a wider exchange of views and detail of practices would be of interest.

7.3.2   The steel feedstock is not flat or not stress-free

Here, if non-flat strip (Figure 7.4) is fed to presses, inevitably non-flat laminations will appear. Further if flat strip is fed to presses when it has within it a system of balanced locked-in stress, punching may release these stresses to promote distortion of the intended component shape, e.g. round holes emerge as ellipses. Many of the above stress problems are resolved when laminations are pressed, or annealed and pressed into complete core stacks, but not all.

Frustratingly it is found that identical steel will give excellent results on one press regime but not on another and it is important to build experience of what factors to specify (if specifiable) so that consistent long-term good results arise. A definitive paper on this topic is still awaited from the stamping world.

Ideally a good punching operation should work well with any good feedstock, and good steel should perform well in any adequate press operation. Mention has already been made of the disparity of punching practice between UK/USA and the Far East (Chapter 4). It is hoped that in the coming years both sets of operators will find it convenient to become familiar with each other's practices.

7.4   Core assembly

Stacks of laminations are usually built into whole stators before windings are applied and the stacks mounted into motors. Possible fixing methods are:

(1) Cleating
(2) Bolting
(3) Riveting

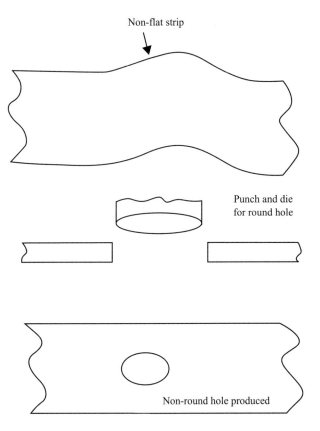

Figure 7.4   Effect of non-flat strip

(4)   Gluing
(5)   Interlocking
(6)   Laser welding in the stamping press
(7)   Electric bead welding
(8)   Die casting of rotors.

(1)   Figure 7.5 illustrates cleating as one of these techniques. Cleats avoid high temperature damage and if well designed are a relatively gentle means of core securing.
(2)   Bolting involves the production of extra holes and is now less used. It can make stators demountable, facilitating rewinding.
(3)   Riveting involves heavy cold work on the ends of a long rivet bar, but if done by hydraulics can be made to apply controllable stress only. If tooth burr is severe rivets can abet long-range short-circuit currents.
(4)   Gluing involves the use of an on-steel coating which when heat-treated consolidates a lamination stack into a firm block. It is important to choose an adhesive

Figure 7.5   Lamination stator stack

which does not apply severe stress to the steel when curing. A comparative study of stress produced by adhesives is needed with frequent updating as new adhesives appear.

(5)   Interlocking is a process in which a punched depression in one strip is filled by a protuberance from the strip above (Figure 7.6). If the depression and pips are correctly designed they interlock under squeezing force to give a secure stack. Figure 7.6 also shows the pattern of metal deformation and grain development which takes place. For interlocking to be consistently successful appropriate mechanical properties of steel must be matched with the relevant interlock tooling.

(6)   Laminations can be welded together at the moment they are produced in a press (by use of a laser) or by electric bead welding as a made-up stack.

(7)   If coatings are used which evolve much gas when heated, then the bead in bead welding can show blowholes; this is undesired by the end-user. The speed of welding bead formation also influences the incidence of blowholes. Experimentation to match welder feed rate and coating properties is needed whenever changes need to be made to steel or coating.

(8)   Die casting of rotors: when rotors receive squirrel cage bars in the form of die cast aluminium the process also acts as a securing system for the rotor laminations. Although at modest slip the rotor current frequency is low it is still useful to aim for interlaminar insulation to restrain eddy currents. The rise in the use of power electronics for motor speed control means that there will be an increasing exposure of rotor stacks to higher-frequency magnetic fields when operating within the normal speed range.

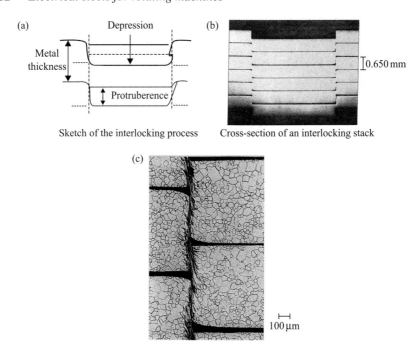

(a) Depression

Metal thickness

Protruberence

Sketch of the interlocking process

(b) ⊺0.650 mm

Cross-section of an interlocking stack

(c) 100 μm

Figure 7.6   Details of the interlock process

Inevitably there is a tension of interest between cost and effectiveness of core assembly methods. The two most detrimental features of core building are discussed in the next two subsections.

7.4.1   Production of interlaminar short-circuits leading to raised eddy currents and high losses

It takes two short-circuits to complete an eddy current loop, and a weld bead down the outside of a stator stack amounts to one short-circuit line in what would appear to be a safe place. However, if burrs formed on the tooth parts of stampings produce short-circuits then these complement the weld bead current paths and raised losses are inevitable.

If weld bead securing is used it should be complemented with specially careful stamping or post-stamp deburring as well as an insulating coating to 'hide' moderate burrs. As will be clear from Ref. [5.1] eddy losses in machines of moderate size seldom rise significantly due to ohmic conduction through coating layers; rather, this is due to the intercrushing of bare metal crystals at the site of burrs providing easy electron paths. Equivalently, if rotor laminations are forced onto a mandrel as part of the securing process (shorting them to the mandrel) then extra care must be exercised in any machining of rotor surfaces to ensure that the turning tool does not smear metal across lamination faces to give a second short-circuit. Although rotor currents may be of relatively low frequency, harmonic components exist and short-circuits are unwelcome.

7.4.2 Production of stress

It is well known that most stresses cause deterioration of magnetic performance. The only exception is mild tensile stress in the direction of magnetisation. Stress can arise from stamping, and in the case of semi-finished steel this is relieved by the final annealing treatment. However, the mode of securing stator lamination stacks in motor casings merits attention. Sometimes stator stacks may be quality tested before cases are fitted, and if this is directly after annealing as a stack it will give a very good result.

If stator stacks are hydraulically pressed into fabricated or cast casings the stress arising can be very great. If cases are heated to facilitate core insertion then cooling of the case will exert severe radial stresses on the core steel. It is true that the thermal contact with the case is of lower impedance when tightly gripped (as compared to a glued-in fixture), but the magnetic permeability of the steel can be greatly deteriorated (higher VA demands) if stress is high. Unfortunately the better-quality steels are more sensitive to stress so that the cost of a superior core metal may be wasted if it is poorly secured.

It can be seen from Figure 10.45 (Chapter 10) that the processes applied to steel after receipt by a motor builder can lead to widely fluctuating magnetic performance. As the legislation governing machine efficiency ranges grips more tightly it will become cost-effective to value-engineer every process of steel handling and use to ensure that the good qualities inherent in the purchased material are carried through to the final machine. These handling features are the preserve of the machine builder, and steel producers can be aware of material responses to abuse but cannot influence what is done, especially if the market climate presses heavily on anything which raises a machine's first cost.

Chapter 8
High-frequency applications

Solid steels have been, and in the main still are, the medium with which transformer cores are made and power converted from mechanical power into electricity and vice versa. In rotating machines the operating frequency of electrical systems ranged from $16\frac{2}{3}$ Hz for some traction up to 50 or 60 Hz for widespread industrial and domestic power usage. Direct current has been largely excluded because of the inability of high power levels to be economically transmitted over long distances at voltages appropriate for consumption but it was often rectified last for traction, e.g. for trams.

The frequencies involved ranged down towards $16\frac{2}{3}$ Hz to fit in with traction motor needs and seldom went above 60 Hz because core losses and system reactance became problematical. Raising frequencies allowed smaller and lighter cores to transmit the same power flows, but led to reduced efficiencies via raised losses.

With the advent of power electronic circuitry, switching frequencies on power conversion systems are becoming ever higher and are approaching 20 kHz plus. Transformers and inductors are essentially used in the conversion path for voltage stepping (up or down) for isolation and for ripple smoothing, respectively. These devices require careful consideration in terms of core loss and copper loss so that temperature rise can be kept within operating specifications particularly for traction applications where allowances for space, weight and cooling are very limited.

In the special case of aircraft and some other limited applications where size and weight of rotating machines are more important than absolute efficiency 400 Hz and up to some 2 kHz have been used. Further, cores with the highest temperature capability could be used involving cobalt–iron alloys where cost was not a prime determinant of choice.

8.1   New emerging applications

While the same factors influencing the choice of 50 or 60 Hz largely still apply, so that these will long remain, some important areas of higher-frequency usage are emerging.

8.1.1   Motor control

There are two main reasons for higher frequencies appearing in motors: simple speed control and high-energy packages.

8.1.1.1   Simple speed control

For many years the three-phase induction motor has been a staple provider of effort to drive machinery. Only limited speed control had usually been on offer involving pole changing or, on small machines, voltage reductions.

Cycloconverter variable voltage/frequency schemes tended to be expensive and inflexible. Modern computer control of industrial systems requires smooth speed control. Usually the speed ranges available were downwards from the 50/60 Hz synchronous rpm and some small percentage upwards.

It would appear that a relatively limited variation in excitation frequency would not place great demands on the core steel. This will be looked at in more detail later. Reducing rpm reduces power output for the same torque. This is acceptable where the requirement of the driven system fits in with this pattern.

8.1.1.2   Higher-energy package

It may be stated by way of reminder that the force between current-carrying circuits is proportional to $B^2 A$ where A is area and B the ruling intensity of magnetisation. This is relatable to torque in a motor and since force $\times$ distance $=$ work, and work/sec $=$ power, the power output from a machine can be raised in proportion to rpm if torque is maintained.

$$\text{Force} \propto B^2 A$$

$$\text{Work} \propto B^2 A \times S \quad \text{where } S = 2\pi r$$

$$\text{Power} \propto \frac{B^2 A \times S}{\text{time}}$$

Notable rises in rpm via raised supply frequency permit more power to be converted through a machine of given size and weight. However radical rises in frequency entail greatly raised core losses (eddy current losses rise approximately as frequency squared) and falling effective permeability of the core metal (as frequency rises flux penetration becomes incomplete). The reduced effective permeability increases magnetising current demands and lowers torque as the effective $\hat{B}$ falls.

Adaptation of the core metal to these altered requirements can go some way towards postponing the onset of reduced power and lowered efficiency which become inevitable at the highest frequencies.

8.1.2   Alternators

A new direction in power generation involves the production of small alternators able to be directly driven from gas turbines and reaching towards 100,000 rpm. The use of permanent magnet rotors can be considered for this application. The power developed is in the kHz range and requires treatment before feeding normal equipment. Polyphase rectification and inversion is practicable. Output frequency, voltage and regulation can be controlled by feedback in the rectifier–inverter system. In this application the considerations are largely those of raised frequency and are much simpler than may be expected with motor control. There is a big foment of activity in this area of work.

8.1.3 Magnetic bearings

While condition monitoring of pressure-fed oil-lubricated bearings gives good protection, a class of applications has emerged in which very low friction and very reliable operation is required up to high rpm values. This need has been addressed by the use of magnetic bearings. It is beyond the scope of this book to examine all the theory of operation of such bearings. It is sufficient to indicate that the machine shaft is maintained in position by a balance of magnetic forces. Both permanent magnets and AC systems may be involved. Essentially the shaft position is maintained via appropriate sensors and control signals used to control the current fed to electromagnets round the shaft so that its position is accurately maintained (Figure 8.1).

Clearly for high rpm values the response time must be in the submillisecond range. Electromagnets able to deliver appropriate forces rapidly must be able to respond quickly. This requires that the magnet current frequencies are high. Core metal required for the duty must be very thin for rapid flux penetration, and of high permeability.

8.1.4 Traction

There are many patterns of power handling in traction systems. Rectification and re-inversion may be required on board a locomotive. Heavy-duty battery charging may require smoothing and interphase reactors. In general traction plant has to be very robust and use air (often forced) rather than oil cooling to minimise fire risk and to reduce size and weight.

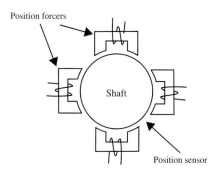

Figure 8.1   Magnetic bearing

Environmental protection of tractive devices has to be of a high order and noise levels must be kept within acceptable limits. The use of air cooling and core metal which is magnetostrictive make this demand for low noise difficult to achieve. Copper losses can be reduced by the use of Litz wire (a multistranded conductor of low HF resistance) to wind inductors and the like where high-frequency operation is required.

8.2   Available materials

Against a background of rising frequency the following materials can be considered.

8.2.1   Conventional electrical steels

These fall in the thickness range 0.65–0.23 mm and have resistivities between some 12 and 50 μΩ cm (depending on the percentage of silicon added). As frequencies rise to 400 Hz and upwards both the hysteresis and eddy current losses rise, the latter the more rapidly. Eventually core losses become so high that operating inductions must be cut back to $\hat{B} = 1.0\,\text{T}$ or less (Figure 8.2). This leads to a need for increased core cross-sections and consequent raised copper losses (the copper has to be longer to encircle a larger core).

In general eddy currents rise with frequency ($V_{\text{RMS}} = 4.4\hat{B}nfA$) since the emf driving eddy currents rises as f^1 and thus watts as f^2 since watts $= V^2/R$. Eddy loss will rise with $(B_{\text{max}})^2$ for similar reasons. Thus

$$\text{eddy loss} \propto \frac{K\hat{B}^2 f^2 t^2}{\rho}$$

Hysteresis loss represented by the number of hysteresis loops traversed per second rises with f^1 and B^{K_2}.

The long-used Steinmetz equation indicates that hysteresis $= K_1 \times (\hat{B})^{K_2} f$ where K_1 relates to the particular material and K_2 is 1.6. K_2 has not been well-charted for modern materials and more work could be done on this.

The developmental directions likely to be adopted for conventional steels are:

(a) *Make thinner:* Steel thickness may be reduced towards 0.05 mm. This radically raises the cost of production but preserves high-field permeabilty. Flux penetration can be substantially complete, avoiding severe skin effects. Space factor (percentage of space occupied by the metal) suffers and falls from 98–99 per cent into the lower 90s. Processing cost increases due to the need for more lamination cutting and a longer assembly time. These increases are unfavourable and drive up the production costs of cores.

(b) *Make more alloyed:* Increased silicon content helps to reduce eddy currents but as the amount used rises above 3% rollability falls and the attainable high-field permeability falls due to iron dilution. If $6\frac{1}{2}\%$ silicon could be used the magnetostriction falls to a low level at this alloy level. Silicon at $6\frac{1}{2}\%$ has been obtained in silicon steel by NKK using a diffusion process to raise levels from

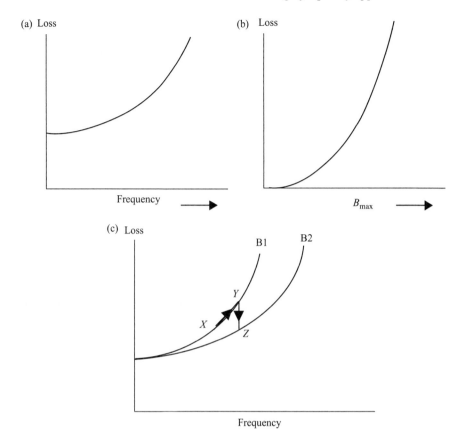

Figure 8.2   *Rise in power loss with*

    a   frequency and

    b B_{max}

    c   If X is an initial design point Y shows the increased loss that results if frequency is increased but induction kept the same at B1. Z shows the reduction in flux density, to B2, that is required to keep core losses at the original level. If this is done the reduced flux gives less torque so mechanical output falls and the horsepower gain arising from increased RPM is lost

3%. This in-diffusion is applied after rolling. This is an expensive process but delivers low magnetostriction values.

(c) *Reduce magnetostriction:* It is worth reducing magnetostriction if possible as 'whine' in traction devices is severe. If the path length in metal coincides with a resonant length machine vibration is greatly increased so that good designs avoid this situation.

(d)   *Control of steel texture:* A lot of research is being applied to texture modifications so that a product has an optimum grain size to deliver minimal high-frequency losses [8.1]. Very small grains deliver high hysteresis losses but beyond a certain size grains have domain walls at a wider spacing such that the extra losses associated with their required rapid movement are disadvantageous. At 50/60 Hz the grain size for transformer steels should not exceed some 5–7 mm wide, whereas for 400 Hz upwards grains of size below 100 microns have advantages.

Micro-eddy-currents are associated with the motion of a domain wall. These micro-eddy-currents are generated as the magnetisation vectors sweep round in the metal lattice. Consequently a domain wall is a dissipative entity. The faster it moves the more dissipative it is. The energy conditions within a grain dictate that as grain size increases domain wall spacing increases so that walls have to move faster to reverse the magnetisation in the available time. So while it is best to keep grains large to minimise pinning effects at grain boundaries, it is advantageous to keep them to less than a size where high-speed walls overly increase losses.

For motor steels of a non-oriented variety it has always been appropriate to grow grains as large as practicable as the metallurgy of production could not give grains large enough to have unprofitably large domain wall spacings at the frequency employed. For high-frequency operation, however, this open-ended aim has to be modified so that an optimum size is chosen. Particularly the challenge is to combine an optimal grain size with appropriate texture so that the properties of the material are substantially the same in all directions in the plane of the sheet while avoiding the appearance of many crystal cube body diagonal directions in the plane of the sheet.

8.2.2   Amorphous metals

In amorphous material the metal acts as one huge dislocation net so that a domain wall is in the same energy state whatever its position. This avoids the specific pinning point difficulty found in crystalline materials and allows domain walls to glide smoothly and to show a very low hysteresis loss. However, stress or mechanical damage produces high hysteresis. Eddy current loss remains low because the resistivity of the metal is high and the thickness is necessarily very low due to the production methods. Efforts are being made to produce amorphous metals in bulk form, but there is a long way to go before commercial products may be expected.

Amorphous metals have the advantage of being very thin, e.g. 25 microns, while having a high resistivity, e.g. $100 + \Omega\,m \times 10^8$. This leads to very low eddy current losses. However the high alloy content limits high-field operation and magnetostriction noise is quite high. Amorphous metal is awkward to handle and requires protection from stress. Microcrystalline and nanocrystalline (not amorphous) versions of spin cast metal are coming into wider use. Operation at frequencies up to 20 kHz seems practicable. These materials offer only a limited maximum width, e.g. 30 cm, hence applications tend to be limited to low-power devices.

8.2.3 Cobalt irons

These materials have excellent properties, such as high saturation induction, up to 2.4 T and Curie points up to 900+ °C, and dominate niche markets where high J_{sat} and high Curie points are essential. Their high cost, however, makes them unattractive for widespread industrial use. Applications range from aircraft generators to submarine drives.

8.2.4 Composites

Composite materials are becoming more and more readily available in a variety of forms and grades. Essentially they consist of an agglomeration of ferromagnetic particles (iron or silicon–iron) compacted in such a way that each particle is insulated from its neighbour either by a natural surface oxide or an applied bonding/insulating coating.

These materials have the benefit of low eddy current losses arising from the insulated particle format but the interparticle reluctance requires heavier magnetising fields to produce high inductions. Permeabilities of the order of hundreds are more usual and are much lower than the 1000/3000 for low alloy silicon iron. However the ability of such composites to be formed into near net shapes and to have the same permeability in all directions means that new challenges and new opportunities are presented to machine designers to exploit these features [8.2] [8.3].

8.2.5 Ferrites

If frequencies are pressed to very high levels, ferrite materials become attractive, but their reduced J_{sat} figures and mechanical fragility mean that ferrites only become attractive in a frequency range well above the capability of most steels.

Although amorphous metals, cobalt–irons, composites and ferrites have been considered here in the light of high-frequency needs, they can all perform at low and zero frequency if other circumstances make this appropriate. This is persued in Chapter 12.

8.3 Impact of new demands

The need to operate motors and generators at the higher frequencies associated with improved speed control and energy throughput for a given size and weight is demanding enough, but further difficulties arise. To create higher frequencies (for example) in motor speed control systems, pulse width modulation (PWM) techniques are generally employed. This technique uses a carrier frequency well above the frequency of the supply it is desired to deliver to a motor and, by operating on the mark-space ratio at which the carrier calls energy from a nominally DC supply, a shaped waveform of the desired frequency is built up. A variety of strategies [8.4] are available for motor supply shaping.

It is necessary to switch semiconductors very rapidly from fully 'on' to fully 'off'. Transition states are very dissipative and damaging to the switched devices and to system efficiency. Consequently the generated waveforms contain a great many fast edges. Fourier transforms of these edges do of course show the presence of a wide spectrum of harmonics.

These raised frequencies are delivered to the motor. Clearly the reactance of the motor windings prevents exact response to much of the most rapidly applied voltage changes, but nevertheless the motor core experiences many minor loop excursions and eddy current losses at frequencies above the desired fundamental. In effect the motor is being used as a harmonic filter and smoothing device.

A variety of papers have been published attempting to extract the specific effect of PWM operation on motor losses. Their findings conflict and extra losses are considered to range between 'very little' and 20+ per cent. Clearly the effects of PWM impact on the response of materials other than conventional solid steel, some more than others.

8.3.1   Efficiency

Conventional induction motors are coming under increasing pressure to be energy efficient. For the 10–100 kW range efficiencies of some 90% are usual. Great efforts on the part of motor producers and steel-makers have combined to raise this efficiency by some 3% without adding extra cost. At the same time European regulatory bodies are framing strict new achievement clauses for motors that will outlaw those of reduced efficiency. All this Green effort may contrast with the possible behaviour of speed-controlled motors and emerging high-speed machines.

Very often pumps and fans driven by induction motors have been subjected to mechanical throttling to secure reduced throughput (Figure 8.3). This is terribly wasteful and speed-control of the motor is a much preferable alternative. So great are the gains from speed-control rather than mechanical throttling that inefficiencies related to PWM, when it is used, appear insignificant. However the searching light of regulation which applies to ordinary motors may at some stage be brought to bear on speed-controlled machines.

High rpm machines aimed at maximising the output kilowattage of small devices may or may not attract the attention of regulatory intent. Very high speed machines may not be operating steel in low loss regimes but 'system' efficiency may be acceptable. Table 8.1 illustrates the point.

8.3.2   Shaping the demands placed on core steel

So far demands placed on steel suppliers have arrived with little or no consultation from the power electronics world. The rapid strides made in computer systems have seen the progressive fall in the cost of computer systems and at least computing power comes almost free compared to power-handling hardware.

Under these circumstances it would seem reasonable to challenge the power electronics world to produce supplies tailored, as far as possible, to the efficient operation of core steel. It would seem reasonable to request that supplies of variable voltage

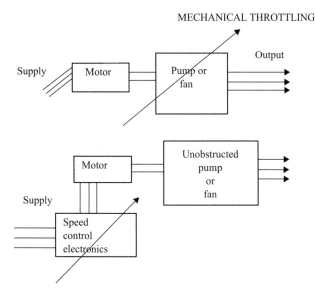

Figure 8.3   Control of flow by motor speed rather than by throttling

Table 8.1

Frequency of operation	Percentage efficiency	kW O/P	Power loss kW	rpm
A 50 Hz	90	100	10	3000
B 100 Hz	80	200	40	6000
C 100 Hz	87	150	20	6000

NB: Figures are notional only.

A: Represents normal full load use.

B: Doubled frequency, doubled speed, if torque is held constant (double input volts needed). but core steel loss rises approximately as f^2, so iron loss quadruples; also copper loss and windage, etc. rise. Efficiency falls to 80% but doubled output is secured. This version would not be chosen, see Figure 8.2.

C: Doubled frequency but operating induction reduced to $\frac{3}{4}$ of that in A. Loss goes up less, about $\times 4 \times 1/2$, i.e. about $\times 2$. Efficiency is quite good but output is only up $\times 2$.

and frequency be supplied in good sine waveform and free from high-frequency components.

Work done to study the domain response of electrical steels has shown that waveforms other than sinusoidal can have the lowest losses. Either a ramp of constant dB/dt or a relative of this waveform minimises losses related to rapid domain wall movement [8.5].

Could not the power electronics world produce this also? Power electronics always needs short-term energy storage so that rapid fast edges can be blunted and moulded into a smooth fundamental-only waveform. Studies of the progressive switching of reactances could provide appropriate insights. Where motor stator laminations are punched 'square' even for round machines, some unused metal appears at corners. Suitable holes, if wound, could provide useful reactances. Stator windings themselves can act as inductive energy stores if this is included in the design.

It would be useful if in-depth conversations could occur between power electronic/computing experts, motor designers and steel-makers so that system efficiencies could be optimised without waiting to be pressurised years hence by regulatory bodies. It will be important to actively design new machines for raised frequencies rather than merely applying raised input frequencies to standard motors.

8.3.3   Testing

It has long been the practice to characterise steel for power frequency use by loss and permeability tests applied under sine wave conditions. This may be a quite unrealistic method of assessing steel for use in conditions of high harmonic content at high frequency.

Perhaps a square-wave applied voltage or some other regime would be useful at 1, 2, 3 kHz. It would be very beneficial if the foundations of a test regime could be agreed before a plethora of ad hoc systems are developed in disparate parts of the industry, requiring difficult rationalisation at a later stage (Figure 8.4).

When soft magnetic materials are used as core metal for transformers, motors and generators of the type employed over the last 100 years the waveform of flux encountered, whilst not exactly sinusoidal, has been sufficiently close to it for sine operation to be adopted as a reference condition for commercial testing. Such testing is for the purpose of categorising metal for purchase and sale rather than the production of the most precise figures attainable. Refinement of testing has consisted of system development to allow sine conditions to be more easily and cheaply attained, e.g. negative feedback amplifiers.

Now	sine B	50/60 Hz
Likely	sine B	50–2000 Hz
	sine H	kHz–nkHz
	Controlled rise time square-wave – to be worked out.	

Figure 8.4   Test methods

The emergence of inexpensive inverters has led to a tendency to operate motors over a wide range of frequencies with two motivations:

(a) To have access to speed control so that pumps, fans, etc. can be flow-controlled by induction motor speed adjustment rather than by throttling the pump or fan (grossly wasteful);
(b) Access to speeds higher than that of 50 Hz supply synchronous speed so that more horsepower (kW) per unit size/wt can be obtained from a given package. This may involve supply frequencies into the kHz range.

8.3.4   Types of inverter

Unfortunately the attraction of inexpensive inverters used to provide high frequency and variable frequency excitation has led to the choice of the cheapest rather than the most sophisticated. Motor supply waveforms are being generated by semiconductor switching (semiconductors perform poorly in other than 'full-on' or 'full-off' modes) with little regard for the effect of the fast edges of the chopped waveforms used. The technique of pulse width modulation exploits the flexibility of semiconductor operation to assemble crude approximations to variable and higher-frequency supplies with no regard to the ill-effects of this.

8.3.5   Ill-effects of pulse waveforms

(a) A Fourier analysis of fast edges shows a range of strong harmonics associated with switching edges. These create strong eddy currents in machine core metal and produce increased machine losses. Regrettably the fascination with system efficiency arising from pump/fan control so overshadows lowered motor efficiency due to PWM that it receives little attention. When large motors use higher-voltage supplies – several kV – the semiconductors in the inverters must voltage-share in chains. The judicious use of capacitors bridging string members, and appropriate algorithms of device firing-order enables good sine waves to be generated. This could be done for smaller machines, but it costs a bit more so that little effort is made to generate machine-friendly waveforms.
(b) The generation of harmonic contamination of mains supply lines leads to increased losses in distribution transformers which may need de-rating to cope (losses in copper conductors rise also). It is hoped that legislation will emerge to restrain mains pollution.

8.3.6   Testing

Clearly sine wave testing does not give a full picture of material behaviour under pulse conditions so that choice for duty becomes difficult. Some test is required which considers pulse response. Every motor situation is different so a myriad waveshapes and frequencies apply. It seems likely that if an arbitrary choice of waveform is made and adhered to, e.g. square-wave applied voltage with defined rise times and fall times, it will not represent every situation but provide a better guide than sine operation. Left unregulated many informal test systems will arise and be very difficult to standardise later.

It is well known that minimum power loss in core steel can be attained by the use of waveforms other than sinusoidal B. For example, constant dB/dt, ramp-up–ramp-down waveforms are good and shapes could be tailored for minimum loss. 'Square wave testing' is a loose term. Does it mean 'square wave applied voltage from a very good regulation source'? or 'such a supply that flux in the metal follows a near square-wave pattern'? or 'square-wave magnetising current'?

Due to the circuit inductances (which vary as permeability varies during changing magnetisation) a square applied voltage can lead to an approximate ramp-shaped B waveform. Overall it seems that the parameters to be defined are:

(1) Frequencies – spot values
(2) B_{max} values
(3) Applied waveshape from a source of what regulation? (Inductance/resistance of magnetising coils, etc.)
(4) Expectation of current waveforms
(5) Constraints on wave rise–fall times
(6) Definitions of the loss and permeability which are to be deduced from the measuring system
(7) Sample and test rig forms.

8.3.7   Development of tests

To move forward in a useful way requires co-ordination between:

(a) Core metal producers
(b) Machine designers
(c) Inverter designers and semiconductor developers
(d) Legislative authorities
(e) Researchers into machine operation conditions
(f) Standards bodies

Open questions are:

• Will machine-friendly low-pollution systems gain wide use due to improved abilities or the force of legislation?
• Will lowest first cost always rule?
• If the IEC sets up a discussion group of appropriate persons the way forward may become clearer and it may be possible to influence the course taken.

8.4   Review and outlook

The trend towards devices operating at higher frequencies and higher harmonic contents is providing a challenge for conventional solid steels. The likely reaction is for these to be provided at thinner thicknesses, perhaps with polished surfaces so that the space factor is maximised. Very thin but effective surface insulation is needed since high frequencies imply comparatively high interlaminar voltages. A judicial choice of alloy level can give increased resistivity while high-field permeability is held at a good level.

The metallurgical challenge lies with optimisation of grain size and material final texture. Although the competition from alternative materials may increase, the benefits of solid steel have good prospects of keeping this core metal in the forefront of markets.

It is now a good time for cooperation between power electronics experts, steelmakers, motor designers and transformer designers to be developed so that the ever-rising tide of Green demands can be appropriately met.

References

8.1. TANAKA, T. *et al.*: 'Effect of metallurgical factors on the magnetic properties of non-oriented electrical steels under PWM excitation', *J. Mag. Mat.* 1994, **133**, pp. 201–4.
8.2. PERSSON, M.: Seminar on magnetic materials based on powder metallurgy, 4 February 1999, Inst. of Physics, London.
8.3. KRAUSE, R. F., BULARZIC, J. H. and KOKAL, H. R.: 'Advances in a soft magnetic material for AC and DC applications', seminar on magnetic materials based on powder metallurgy, 4 February 1999, Inst. of Physics, London.
8.4. BOGLIETTI, A., FERRARIS, P., LAZZARI, M., and PROFUMO, F.: 'Iron losses in magnetic materials with six step and PWM inverter supply', *IEEE Trans. Mag.*, 1991, **6**, pp. 5334–6.
8.5. BECKLEY, P. and THOMPSON, J. E.: 'Influence of inclusions on domain wall motion and power loss', *Proc. IEE* 1970, **117**, pp. 2194–200.

Chapter 9

Finite element design methods – availability of suitable data

9.1   Introduction

The design of a rotating machine requires that the outcome of deploying magneto-motive force from coils or permanent magnets is predictable so that figures for torque and other aspects of machine performance can be produced. Traditionally a wealth of experience built up by designers incorporating the known properties of electrical steels and other materials has been incorporated into computer programs and expert systems to aid design. This process works well but may not be best placed to most advantageously take account of the properties of new materials or to devise routes towards new machine performance demands. The growing use of permanent magnets and operational current waveforms synthesised from multipulse inverters poses new design challenges.

The physical phenomena associated with electromagnetism can be described in terms of a set of partial differential equations which describe the structure and action of a machine. Several properties of the materials make explicit solution of these equations difficult by the usual methods.

Among these are:

- The $B–H$ characteristic of magnetic materials is non-linear and exhibits hysteresis.
- The magnetic permeability of steel varies with direction in three dimensions.
- Discontinuities are normal in machines, such as air gaps, high-reluctance interfaces and rapid changes in cross-sectional area.
- Demagnetising factors at discontinuities are significant.
- The properties of magnetic materials are frequency (and thus time) dependent, so that phenomena such as skin effects distort the usual flux magnifying properties of electrical steel.

One of the tools for the solution of the electromagnetic problems is the finite element method (FEM). In this chapter we will discuss the main points in the application of the FEM to electromagnetic design including the solving of equations and their implementation. Those who seek deeper understanding of the FEM should consult some of the work listed in the Bibliography section (p. 307).

9.2   The finite element method

The finite element method is based on the concept of dividing the original problem domain into a group of sub-domains, the 'elements', and applying a numerical formulation based on interpolation theory to the elements. A numerical solution is then sought with respect to some optimal criterion.

The finite element method was first applied to solve structural analysis problems as early as the 1950s, also the method was applied to heat transfer and fluid flow problems. In 1970, an article by Silvester and Chari entitled 'Finite element solution of saturable magnetic field problems', in which they proposed a formulation capable of dealing with complex geometry and the problem of magnetic non-linearity, signalled the beginning of a new era in the field of applied electromagnetism [9.1]. They made major contributions to the development of the method which is so widely applied today in electrical engineering.

The finite element method is in fact a numerical technique for solving large-scale, complex problems, using a simple and flexible data structure. In order to discretise a problem region, it is necessary to choose elements of a given shape. Several element shapes are in use and the principle ones are: the triangle, the quadrilateral, and curvilinear shapes [9.2]. Elements are defined in terms of their shape and the order of polynomial interpolation of the trial function (the function that describes the variation of the solution parameter within an element in terms of the element's nodal values). Figure 9.1 illustrates the formation of a net and the effect of varying the mesh fineness.

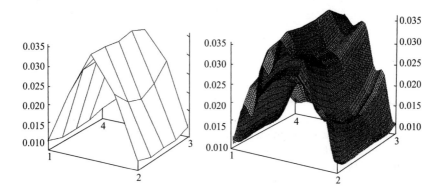

Figure 9.1   Effect of varying fineness of mesh

9.2.1   Finite element formulations

Instead of expressing the laws of electromagnetism in terms of electric and magnetic fields, it turns out that it is often more convenient to express the theory in terms of potentials which are related to the field equations by curl ($\nabla \times$) or gradient ($\nabla \cdot$).

In any 3-dimension, the vector $\bar{b}$ has the property that

$$\nabla \cdot (\nabla \times \bar{b}) = 0 \qquad (9.1)$$

and the Maxwell equations govern the fields in electromagnetic devices are

$$\nabla \cdot \bar{B} = 0 \qquad (9.2)$$

$$\nabla \times \bar{H} = \bar{J} \qquad (9.3)$$

where $\bar{B}$ is the magnetic flux density, $\bar{H}$ is the magnetic field intensity and $\bar{J}$ is the current density.

Therefore, the magnetic vector potential $\bar{A}$, is defined as

$$\nabla \times \bar{A} = \bar{B} \qquad (9.4)$$

By the analogy to the electrostatic field in which the electric field $\bar{E}$ is related to the applied voltage V as ($\bar{E} = -\nabla V$), the magnetic scalar potential ψ is defined as

$$\bar{H} = -\nabla \psi \qquad (9.5)$$

Therefore the Maxwell equations could be formulated in terms of the magnetic potentials $\bar{A}$ and ψ, and finding a solution in terms of such potentials allows the calculation of $\bar{H}$ and $\bar{B}$.

Computer programs are used to solve the FEM formulated equations. Although the available computer power has greatly increased, computer power still limits large 3D models. Therefore the FEM formulations are usually optimised towards faster solutions. One common strategy is to split the problem regions up into non-conducting and conducting volumes and use the optimum field variable in each. Often non-conducting regions are modelled using magnetic scalar potentials, this is because scalar variables are very economical as only one variable is required per node, in contrast with vector potential variables which require three.

Various magnetic scalar potentials are in common use, the total scalar ψ is optimal in regions which contain no current and the reduced scalar ϕ allows current to be introduced into magnetic scalar potential regions. The total scalar ψ is defined as in equation (9.5): $\boldsymbol{H}_T = -\nabla \psi$, and the reduced scalar [9.3] ϕ is defined as $\boldsymbol{H}_T = -\nabla \phi + \boldsymbol{H}_S$. Here $\boldsymbol{H}_T$ is the total magnetic field intensity and $\boldsymbol{H}_S$ is the field defined as $\nabla \times \boldsymbol{H}_S = \boldsymbol{J}_S$, where $\boldsymbol{J}_S$ is the source current density.

The basic method has been extended to allow voltage forced conditions [9.4] and to automatically produce cuts for solving multiply connected problems [9.5]. Both scalars give rise to a Laplacian type equation which has to be solved:

$$\nabla \cdot \mu \nabla \psi = 0 \qquad (9.6)$$

Eddy current regions must always be described by a vector variable. The most common in use today is the magnetic vector potential A. This is defined as $\nabla \times A = B$:

$$\nabla \times \left(\frac{1}{\mu} \nabla \times A \right) = -\sigma \left[\frac{\partial A}{\partial t} \right] \tag{9.7}$$

A fourth equation may be obtained by reiterating the requirement that $\nabla \cdot J = 0$, already implicit in (9.7):

$$\nabla \cdot \sigma \left(\frac{\partial A}{\partial t} + \nabla V \right) = 0 \tag{9.8}$$

In the above V is the electric scalar potential and is not always required. The A and ψ variables may be conveniently coupled [9.6] at common interfaces throughout the problem. It should be noted that many other formulations exist. One of the most important is the T–Ω method [9.7]. Here a vector variable T, defined in conducting regions, is coupled to a magnetic scalar defined in non-conducting regions.

One component of the magnetic vector potential A is by far the most common variable in use for 2D problems. This gives rise to a Poisson-type equation as follows:

$$\nabla \times \left(\frac{1}{\mu} \nabla \times A \right) = -\sigma \left[\frac{\partial A}{\partial t} \right] \tag{9.9}$$

9.2.2   Material properties

When a coil is wound on a core of ferromagnetic material the resulting field is greatly magnified and may be described by the relationship $B = \mu_0(H + M)$, where H is the field as derived from Ampère's law and μ_0 the permeability of free space. M is the extra magnetisation arising from ferromagnetism.

The machine designer is interested in B rather than M, and because B may be composed of a complex mix of vectors which are a function of H and directional material properties, it is hard to quantify. Particularly B shows hysteresis and is direction and time dependent.

The relationship between B and H may be approximately realised in the type of curve shown in Figure 9.2. The steep portion of the curve fades into a region of shallower slope leading to saturation.

9.2.3   Magnetisation curves for use in finite element schemes

Usually in finite element schemes only the mean magnetisation curve is used. This means that such effects as hysteresis cannot be described and the B–H curve becomes single valued and monotonic. The equations will be non-linear if the permeability of the iron cannot be considered constant over the range of operation of the device. These non-linear equations are solved numerically using a standard technique such as simple iteration or Newton–Raphson.

The mean magnetisation curve would normally be calculated from a B–H curve, measured in a standard way. The B–H curve itself is rarely in fact used directly. In magnetic scalar equations such as (9.1), the magnetic scalar is defined as $H_T = -\nabla \psi$,

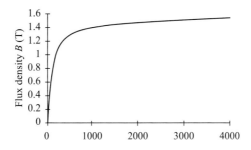

Figure 9.2   B–H characteristic

and the magnetisation information appears as μ in the governing equation. Since the term μ is initially unknown in a non-linear problem, an iterative solution is required. This involves repeated solution of (9.1) and the value of μ is updated at each iteration.

The procedure is to find H from $H = -\nabla \psi$ and to find μ from a graph of μ versus H^2. The Newton–Raphson method requires the slope of the graph of μ versus H^2.

In a similar way, if we are dealing with a governing equation describing the magnetic vector potential, as in (9.9) or (9.7), since the magnetic vector potential is described as $\nabla \times A = B$ and the magnetisation information appears as $1/\mu$ in the governing equation, a graph of $1/\mu$ versus B^2 is required. In addition to this, a graph of the slope of $1/\mu$ versus B^2 is also required if the Newton–Raphson method is used. The smoothness of the various graphs is important for the convergence of the various non-linear solution methods. It can affect the speed of convergence and in extreme cases there will be no convergence at all.

To this end, the magnetisation details would normally be stored in the computer in such a way as to ensure smooth curves. Various curve fitting methods have been used to represent the measured data, including cubic splines (probably the most common method), cubic Hermite polynomials or exponential functions. In addition to this, a good finite element package would allow the user to easily create (or automatically create) curves of μ versus H^2 and $1/\mu$ versus B^2 and their respective derivatives from the measured $B–H$ data.

A good introduction to basic material briefly described here may be obtained from [9.8] or [9.9]. More detail and some practical examples may be found in [9.10].

9.2.4   Hysteresis and iron loss

Some new developments involve the inclusion of hysteresis effects into finite element models; here various different representations have been used including Preisach models [9.11], [9.12] and graphical methods [9.13]. Semi-analytic methods have also been developed [9.14].

Iron loss can be determined in an approximate way by performing a time transient analysis, Fourier analysing the resulting B field in each element and calculating the

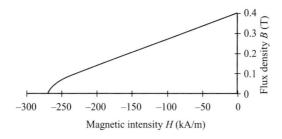

Figure 9.3   Permanent magnet: second quadrant characteristic

loss from manufacturer's iron loss curves. Usually these curves are of loss versus peak **B** and frequency, so interpolation would be required [9.15]. Permanent magnets are used in the second quadrant of the *B–H* plane, see Figure 9.3.

A permanent magnet will operate along the line AB in use (as part of a machine) and may exhibit minor recoil loops. μ_r, the relative permeability which the line expresses, is called the relative recoil permeability.

9.3   Electromagnetic CAD systems

The use of the finite element technique for numerically solving the electromagnetic field equations is nowadays accomplished by fast, powerful and general-purpose software packages. Most of those packages are developed to solve general electromagnetic systems rather than problem oriented programs. These systems are known as electromagnetic CAD systems (CAD being computer-aided design, a terminology that stresses the importance of these systems in the design). Electromagnetic field simulators or simply finite element packages are other popular ways of referring to these systems.

Most of the CAD systems for the numerical analysis of electromagnetic problems are based on the finite element method. The method has proved to be flexible, reliable and effective.

The finite element packages are powerful tools in research, development and design. By using only a personal computer, it is possible to analyse a number of different geometries and operating conditions, without the need of building a physical prototype. Also, the numerical simulation provides, in most cases, reliable and accurate information about the device's behaviour irrespective of geometric complexity and material non-linearity. For example, in permanent magnet motors, the FEM can analyse accurately various shapes and materials. There is no need to calculate reluctances, leakage factors or operating point on the recoil line. The PM demagnetisation curve is input to the FEM package, which can calculate the variation of magnetic flux density throughout the motor. Moreover, the ability to calculate accurately the armature reaction effect and torque variation with rotor position is an important advantage of the FEM over the analytical approach.

9.3.1   General FEM steps

Generally most FEM packages share common steps in constructing electromagnetic models. These steps are as follows:

1. Pre-processing. In the pre-processing stage, the device under consideration needs to be drawn first. It could be drawn by the FEM package itself or it could be imported from other drawing packages. Only active components of the device need be drawn. An active component means one that is part of the magnetic circuit. The actual units of length need to be set out. Material properties need to be assigned to each component.

   Usually a library of various material properties is available to the designer to choose from. The excitation in terms of current in coils in the circuit, number of turns, or permanent magnet nature and direction must be defined at this stage. After that, the defined geometry will be meshed. Usually there is an automatic mesh generator which generates a uniform mesh for the whole geometry. The user may intervene to refine the mesh in the area of interest to increase the accuracy of the solution. Boundary conditions need to be set at this stage in ways which assist solution and minimise the model size.

2. Solving. In this stage the FEM programme starts to automatically solve the formulated field equations implicit in the pre-processing stage. The user needs to choose the appropriate solver for a given design. The solver will be chosen according to the required analysis. There are three main solver types usually used in electromagnetic design.

   (a) Magnetostatics: This is used in analysing the magnetic field in and around specified current distributions or permanent magnets in the presence of magnetic materials, which may be linear or non-linear, isotropic or anisotropic. The solved field values are time independent or are a snap-shot in time-varying fields.

   (b) Time harmonics: This is applicable for devices with sinusoidal excitation and eddy currents, taking skin effects into account. The source and field are assumed to be time harmonic at one specified frequency. Complex phasors are used to represent them.

   (c) Transient solver: This is used to analyse devices with arbitrary shaped current and voltage responses, or to model the effect of transient excitation or motion. It finds the time-varying magnetic field in the presence of material which may be magnetically linear or non-linear.

3. Post-processing. The solver output may be processed to find the required parameters. The user may manipulate the solution to calculate, for example, torque, inductance, etc. Most FEM packages offer graphical representation of the solution, where flux plots or flux density values may be plotted and graduations of colour produced according to the flux levels in different parts of the device.

Based on the steps mentioned before, most of the general purpose FEM packages could be tailored for design and analysis of specific electromagnetic devices like motors and actuators.

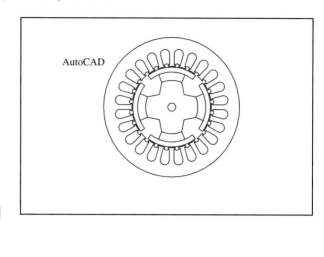

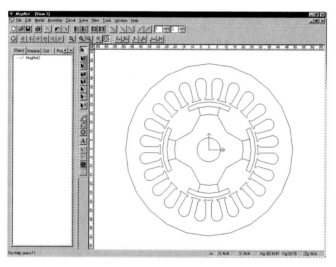

Figure 9.4   CAD translated to FEM damain

Below is given an example of modelling a permanent magnet machine which explores the modelling steps mentioned before (Figure 9.4).

1. Pre-processing

    *   The model geometry could be drawn using AutoCAD and then exported to an FEM package, or it could be drawn directly in the FEM package using its drawing facilities as shown below:
    *   The magnetic materials assigned to each component of the machine, for example the stator laminations, have different magnetic material to that of the rotor.

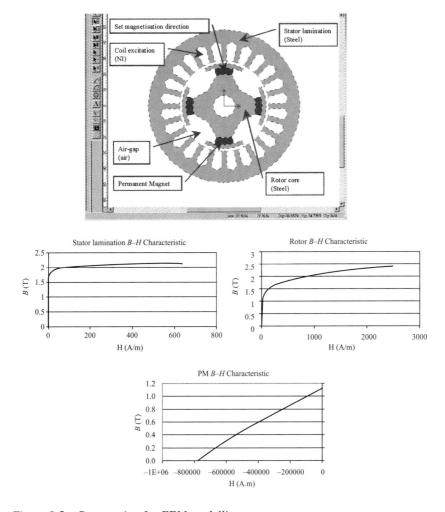

Figure 9.5   Preparation for FEM modelling

Permanent magnet material and its magnetisation direction are set in. The stator winding details and its excitation is also set in (Figure 9.5).

- An automatic mesh generator could be initiated to uniformly mesh the whole model. Refined mesh may be applied in the area of interest. Here, for example, a refined mesh is used near the pole tips to accurately calculate leakage flux (Figure 9.6).

2. Solving
    The FEM package will solve the model defined in the previous step automatically. The user needs to assign the proper solver. In this example, the magnetostatic solver can be used, as flux distribution and magnetisation level is required (Figure 9.5).

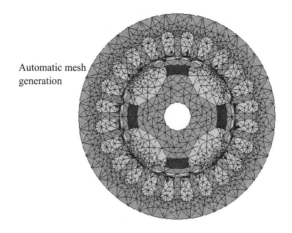

Automatic mesh generation

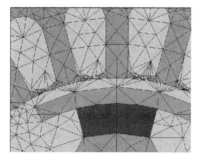

Mesh before refinement

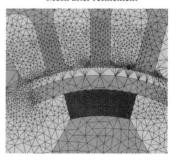

Mesh after refinement

Figure 9.6   Mesh refinement

3. Post-processing
    The solved model result can be accessed in this stage. The flux and flux density values may be plotted over the whole area of the model or in particular parts. The saturation effects may be examined. The parameters of the machine, like inductance, flux linkage or the torque, could be calculated for different saturation levels.

9.4   General comments

The chosen approach to the use of FEM depends to some degree on the size of the engineers' organisation. Only large organisations can afford to employ and maintain an FEM specialist (or more than one) to devise in-house software. Again, putting work out in-toto attracts high fees and may need many attempts to arrive at the desired knowledge. Probably an intermediate approach has much to offer if one individual takes a considerable interest in FEM work.

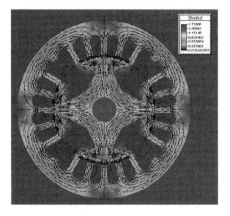

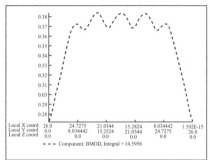

Figure 9.7   Example of FEM solution

Some FEM work may be grant aided from within university departments supporting local industry, especially SMEs. Today when high-level computer languages are in wide use because of their user friendliness, do-it-yourself FEM packages are gaining in popularity. A two or three day course may be an appropriate starting point for would-be FEM users.

To the FEM enthusiast the computer rules and contact with hardware is viewed as condescension. However, ab initio designs conducted by computer alone can fail to incorporate some vital feature of the system. It is often productive to make up a rough model of a rule-of-thumb designed device and note that it operates believably. Then FEM techniques can be used to refine and optimise the design and apply value engineering strictures to produce the most favourable device.

The writer prefers the sequence:

- Idea, rough test, refine and optimise by FEM
- Do everything on computer then build.

Very often components have symmetries such that a solution for one half or one quarter of a device is sufficient, or a solution in one plane may be extended indefinitely in a third direction. See Figure 9.8.

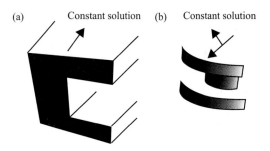

Figure 9.8   *a*   Translational (Cartesian) geometry
          b   Rotational geometry

Structures built from anisotropic electrical steel are notorious for the tensor non-linearity of their properties, for hysteresis and the frequency dependence of their properties. Experience is the best guide of how far approximation may be applied without losing the usefulness of the outcome. For instance, computer results which predict 2.5 T regions in steel under modest applied fields must be suspect, etc.

Many computer solutions represent results in false colour and this greatly aids the subjective interpretation of information. Value engineering may call for more rounded corners, easier-to-punch shapes, etc. Where can metal be safely omitted to allow a bolt hole?

9.5   Outlook for FEM

FEM design can greatly ease the task of getting the best out of a newly developed steel. It can give the metallurgist a view of the most desirable (economic?) metallurgical properties to aim for.

The engineer moving towards a new device can 'test' old and new steels and see how to get the best from any of them. Inevitably steel properties show a spread in production. FEM can map the impact of such spreads on machine behaviour.

With advances in computing power it could become possible to vary the input material property description and the geometrical and time constraints while watching the results in real time rather than awaiting solution by solution.

Source of software package:

Vector Fields Ltd.
24 Bankside, Kidlington, Oxford OX5 1JE, UK
Tel: +44 (0)1865 370151
Fax: +44 (0)1865 370277
Internet: http://www.vectorfields.co.uk/

Infolytica Limited
68 Milton Park
Abingdon, Oxon OX14 4RX, UK
Tel: +44(0)1235 833288
Fax: +44(0)1235 833141
Internet: www.infolytica.com

Laboratory offering developed problem solutions:
Wolfson Centre, School of Engineering,
Cardiff University, The Parade, PO Box 687,
Cardiff CF2 3TD

References

9.1. SILVESTER, P. P. and CHARI, M. V. K.: 'Finite element solution of saturable magnetic field problems', *IEEE Trans. Power Apparatus Syst.*, 1970, **89** (7), pp. 1642–50.

9.2. ZIENKIEWIEZ, O. C.: 'The finite element method in engineering science' (McGraw-Hill, London, 1977).

9.3. SIMKIN, J. and TROWBRIDGE, C. W.: 'On the use of the total scalar potential in the numerical solution of three dimensional magnetostatic problems', *IJNME*, 1979, **14**, pp. 423–40.

9.4. LEONARD, P. J. and RODGER, D.: 'Modelling voltage forced coils using the reduced scalar potential method', *IEEE Trans. Magn.*, 1992, **28** (2), pp. 1615–17.

9.5. LEONARD, P. J., LAI, H. C., HILL-COTTINGHAM, R. J. and RODGER, D.: 'Automatic implementation of cuts in multiply connected magnetic scalar regions for 3D eddy current models', *IEEE Trans. Magn.*, 1993, **29** (2), pp. 1368–71.

9.6. RODGER, D.: 'Finite-element method for calculating power frequency 3-dimensional electromagnetic field distributions', *IEE Proceedings Part A*, 1983, **130** (5), 233–8.

9.7. PRESTON, T. W. and REECE, A. B. J.: 'Solution of 3-dimensional eddy current problems: the T-omega method', *IEEE Trans. Magn.*, 1982, **18** (2), pp. 486–9.

9.8. SILVESTER, P., CABAYAN, H. and BROWNE, B. T.: 'Efficient techniques for finite element analysis of electric machines', *IEEE Trans. Power App. and Syst.*, 1973, **92**, pp. 1274–81.

9.9. SILVESTER, P. P. and FERRARI, R. L.: 'Finite elements for electrical engineers' (Cambridge University Press, Cambridge, 1990).

9.10. PRESTON, T. W. and REECE, A. B. J.: 'Finite element methods in electrical power engineering' (Oxford University Press, Oxford, 2000).

9.11. BERQUIST, A. and ENGDAHL, G.: 'A stress-dependent magnetic preisach model', *IEEE Trans. Magn.*, 1991, **27** (6), pp. 4796–8.

9.12. MAYERGOYZ, I. D.: 'Mathematical models of hysteresis' (Springer-Verlag, 1991).
9.13. LEONARD, P. J., RODGER, D., KARAGULER, T. and COLES, P.: 'Finite element modelling of magnetic hysteresis', *IEEE Trans. Magn.*, 1995, **31** (3), pp. 1801–4.
9.14. HODGON, M. L.: 'Application of a theory of ferro-magnetic hysteresis', *IEEE Trans. Magn.*, 1988, **24** (1), pp. 218–21.
9.15. Applied Electromagnetics Research Group: 'MEGA Commands Manual', Bath University, BA2 7AY, UK.

General reading

HO, S. L. and FU, W. N.: 'Review and future application of finite element method in induction motors', *Electric Machines and Power Systems*, 1998, **26**, pp. 111–25.
LOWTHER, D. A. and SILVESTER, P. P.: 'Computer-aided design in magnetics' (Springer-Verlag, New York, 1986).
NATHAN, I. D. A. and BASTOS, J. P. A.: 'Electromagnetics and calculation of fields' (Springer-Verlag, New York, 1997).
RILEY, K. F.: 'Mathematical methods for the physical sciences' (Cambridge University Press, Cambridge, 1974).
SILVESTER, P. P. and FERRARI, R. L.: 'Finite elements for electrical engineers' (Cambridge University Press, Cambridge, 1990, 2nd edition).
TROWBRIDGE, C. W.: 'An introduction to computer-aided electromagnetic analysis' (Wessex Press, Oxford, 1990).

Chapter 10

Testing and measurement

Electrical steels are critically assessed not only on their electromagnetic properties, but on their mechanical properties also. Additionally aesthetic considerations apply to finished material.

10.1 Magnetic properties

Electrical steels are marketed primarily in terms of power loss performance. Within a power loss grade the permeability of the steel is next in importance. The thickness and price at which a given power loss range is attained is very influential in stimulating demand.

10.1.1 Power loss

As has been indicated in earlier chapters power loss is the energy dissipated during the cycling of a steel's magnetic state at a frequency and peak induction relatable to its end application. While data which applies to a whole range of working inductions and frequencies is important and available, key grade assessments are taken as the performance of steel at 50 or 60 Hz and at $\hat{B} = 1.5$ or 1.7 tesla under conditions of reversing sinusoidal flux. Particularly in some parts of Europe grading at 1.0 tesla was common as a carry-over from the time when the high silicon levels used in hot-rolled sheet made this test induction appropriate.

It is particularly important that material be grade-selected using tests at inductions which are closely related to the induction which will apply in use. A lower loss grade could look better on paper but perform worse in practice if the grade test induction were far removed from the service condition.

10.2 Test methods

Because of the wide diversity of physical forms and magnetic circuits which apply to steel in use it is convenient to select a physical form for a test sample and standardise

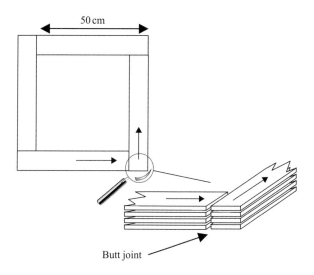

50 cm

Butt joint

Figure 10.1   The 50 cm Epstein test frame: sample layout

this, so far as possible, world-wide. From the turn of the last century efforts have been made to standardise test sample configurations. A brief review of the development route taken is of interest. About 100 years ago the 50 cm Epstein test came into use. The sample layout for this is shown in Figure 10.1. The Epstein test used butt joints between limbs having a square arrangement as shown.

This test avoided samples stressed by curvature but with 50 cm × 3 cm strips cut from parent sheets, cutting stress could be significant unless samples were stress-relief annealed. The butt joints at corners represented significant magnetic reluctance, but with the main limbs 50 cm long this effect was not overwhelmingly bad.

The test frame comprised four sets of windings within which the four sample limbs were placed, each winding having primary (magnetising) and secondary (B sensing) coils. The whole amounts to a small model transformer.

There is no space here to analyse these developments of early loss tests in depth. It suffices to note that all were progressively superseded by the 25 cm ('Baby') Epstein test which emerged around 1936/38 [10.2].

The layout of this 'Baby' 25 cm Epstein test frame is shown in Figure 10.2. The shift to 25 cm Epstein meant that the sample mass was reduced from some 5–10 kg to around 0.5–1 kg when the number of strips in a test was reduced to 24. Meaningful tests can be carried out with as few as eight strips (two per limb) if need be. Figure 10.2 also shows a photograph of two modern Epstein frames (one miniature), and a complete test system.

The Epstein test involved exciting a magnetising winding (700 turns spread over 175 turns per limb of the square) so that the sample attained the desired peak induction. The peak induction was measured by an average sensing rectifier voltmeter connected to the secondary winding of the square (also 700 turns). The voltage from

(a)

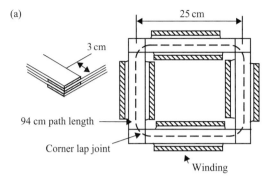

(b)

Complete Epstein test facility.

(c)

25 cm Epstein test frame and miniature research model.

Figure 10.2   *a*   Outline of 25 cm Epstein frame
            b   Epstein test devices

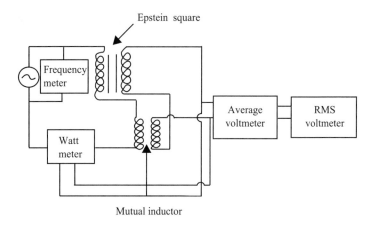

Figure 10.3   Circuit for use of Epstein frame

this secondary winding was used to feed the pressure coil of a dynamometer wattmeter whose current coil was excited by the current in the primary windings of the square.

Figure 10.3 shows the circuit in use. In the early years of such tests sinusoidal induction was not easy to attain, particularly at high inductions. The non-linearity of the iron's *BH* characteristic requires that for sinusoidal flux a very non-sinusoidal excitation current is used. Primary current has to peak sharply as the induction approaches saturation. To achieve a good sinusoidal d*B*/d*t* waveform (and have co-sinusoidal *B*) a source of excitation of very low impedance and excellent regulation, itself accurately sinusoidal, is needed. Recognising that a very good sinusoid is hard to obtain, various devices have been tried to improve matters. The introduction of variable amounts of compensating harmonic current into the excitation circuit has been used.

Not until the availability of negative feedback amplifiers able to force sinusoidal induction was this problem finally overcome. Even so the use of very heavy feedback requiring severe *H* waveform distortion risks instability in the system if gain and phase margins become adverse.

Currently much effort is being applied to digital systems of waveform creation and control so that the main hazards of traditional analogue feedback systems can be avoided.

The final solution coming into use is the application of digital methods for waveform synthesis and control although the ability to sustain sinusoidal flux at higher and higher inductions ceases to be useful when this becomes too far removed from likely service conditions.

Test systems were often excited by use of an alternator driven by a carefully speed-controlled DC motor so that the frequency of test could be closely monitored and amplitude-controlled by variation of the alternator excitation.

If easy access to an oscilloscope was unavailable, the parallel connection of an RMS sensing and an average sensing voltmeter on the secondary winding of the square enabled comparison of these readings to be used to assess the sinusoidalness of the dB/dt waveform. For an exact sinusoid $V_{RMS}/V_{mean} = 1.111$. Departures from this ratio could be noted and a correction applied to the core loss result, in line with agreed formulae [10.3].

It is always useful to have an oscilloscope connected to the secondary circuit of the square so that any anomalies in waveform can be noted and explored.

The key components of an Epstein test system are:

- The test frame
- A source of excitation power
- A wattmeter
- A $\hat{B}$ meter
- An air flux compensation system.

10.2.1 Test frame

The standard variety uses 700 turn primary and 700 turn secondary windings (secondary placed innermost nearest to the sample). As excitation systems have changed over the years, different numbers of turns have been found to be more convenient. For example, semiconductor power amplifiers match better to a lower number of turns.

10.2.2 Excitation source

The local mains power supply fed via a variable autotransformer is a simple but rough and ready source of excitation. The local supply frequency is often uncertain and its waveform imperfect. Locally driven alternators have been largely replaced by solid state power amplifiers coupled with crystal-controlled oscillators. A shift to digitally synthesised and controlled excitation is currently in motion.

10.2.3 Wattmeter

Dynamometer wattmeters depend on the forces between stationary and moving coils interpreted via the modulus of an elastic suspension system. Such wattmeters have been refined by the use of mirror, lamp and scale indication methods, and accuracies in the sub-0.5% region have been attained. This is a high achievement in the face of power factors ranging down to below 0.1.

Dynamometer wattmeters still form the core of traceable power loss measurement. Latterly electronic wattmeters able to develop accurate product signals have become available. They can be robust and accurate.

An interesting, useful and versatile wattmeter has been developed using stochastic–ergodic signal processing to yield power figures from voltage and current infeeds. One such device is well reported in Ref. [10.4].

The increasing use of PCs to offer virtual instrumentation means that power measurement can be provided by an appropriate card in a PC system. This is excellent in principle but the traceability of performance with respect to very low power factors, high peak/mean voltage ratios of waveforms, etc. requires careful attention if results are to be authoritative.

10.2.4 $\hat{B}$ meter

Early forms of $\hat{B}$ meter used an average sensing voltmeter based on a rectifier-fed moving coil voltmeter. This is still a useful, though not a quick and flexible method. Care has to be taken due to such meters often being RMS scaled to meet user demand in electrical engineering applications. The scale has, by appropriate arithmetic, to be re-scaled for correct use.

It will be recalled that the equation $\overline{|V|} = 4\hat{B}nfa$ applies for

$\hat{B}$ = peak induction
n = number of turns
f = frequency
a = cross-sectional area
$\overline{|V|}$ = average rectified voltage.

This formula is waveform independent and reflects the core notion of $\hat{B}$ setting.

Older digital voltmeters used a perfected (diode feedback on an operational amplifier) rectifier and low-pass filter to deliver $\overline{|V|}$ values. These could be RMS scaled when a sine wave was used for the convenience of general use. Under pressure of demand true RMS sensing instruments have been developed. Once expensive, these are now relatively cheap, but often do not offer a true average response as an option. Consequently only expensive versatile meters offer a true average as a refined option.

So the progression has been:

- Pointer instruments + rectifier – inconvenient, delicate
- Early DVMs average sensing – fine
- Recent DVMs, RMS sensing – not useful for $\hat{B}$ measuring
- Expensive do-all versions offering $\overline{|V|}$ and $\tilde{V}$ – fine but expensive.

The demand for cheap $\overline{|V|}$ sensing meters by magneticians is a minute percentage of the whole market so it is not clear if this niche market will ever be well addressed. An accuracy better than 0.2% is desirable. Each 1% error in $\hat{B}$ causes some 2% error in power loss assessment.

10.2.5   *Air flux compensation*

Depending on the detail of coil-former construction and the percentage fill of a coil former by a steel sample, the secondary winding of the Epstein frame may contain a variable amount of air or non-magnetic material. When measurements are made at high inductions the extra emf arising from the 'air and coil former' cross-section makes a significant contribution to secondary coil voltages and errors can arise. At lower inductions the steel sample is so overwhelmingly the source of emf in secondary windings that the error may be small.

It has become policy to connect a mutual inductor as shown in Figure 10.4 so that, when its value is adjusted such that with a frame containing no sample, the net output from the square secondary winding in series with the mutual inductor secondary is zero (for a strong current in the primary windings).

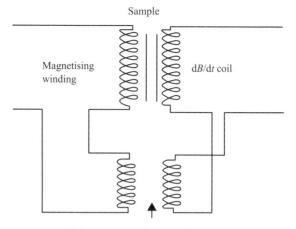

Air flux compensation mutual inductor

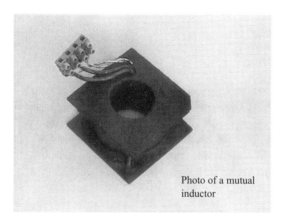

Photo of a mutual inductor

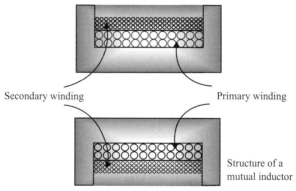

Structure of a mutual inductor

Figure 10.4   *Connections for an air flux compensating mutual inductor: photograph and structure*

This means that the 'B' winding delivers an output relating to steel alone. It also means that the space occupied by the steel when in place is compensated for, so subtracting 1.0 from the relative permeability of the steel.

It is so convenient to use a fully compensated square that this is almost always done. If an exact figure for *B* in the steel, inclusive of the air-equivalent of the space it occupies, is needed this can be adjusted for by calculation from the known cross-sectional area of the steel. Ref. [10.3] considers this matter in some detail.

A fully compensated (empty) square should have a net output with no sample loaded of less than 1% of that appearing with the mutual inductor omitted. If the mutual inductor is located (as is convenient) in the centre of the square care must be taken that the 'extra turn' round it due to the presence of four limb windings in a 'ring' does not confuse the correct setting.

10.3   Path length

As has been indicated the number of primary and secondary turns of the Epstein square can be adjusted to match the supply system and measuring instruments. Clearly, *mutatis mutandis*, arithmetic factoring of the wattmeter reading (of whatever form) has to be applied to allow for the different number of turns.

The cross-sectional area of the steel is most readily determined from its mass and known density. Strips for the Epstein test have to be a least 28 cm long but longer ones (which aid physical loading and unloading of the square) are permitted. If longer ones are used then this must be allowed for when calculating the sample cross-sectional area from the sample mass and density.

The effective path length of a test sample made up of 28 cm strips using double overlapped joints is not a constant. Depending on the nature, particularly permeability, of the steel the configuration of flux at the corners will vary so that effective path lengths ranging from 88 to 106 cm have been noted. To determine absolutely the path length is a complex task involving the use of Epstein frames of differing limb lengths and even then the result only holds for the particular steel used and the induction and frequency employed.

The effective path length is a vital quantity because the percentage of the whole sample mass (as gravimetrically weighed) which is used in the quotient watts/mass to give watts/kg as a result has to be decided upon.

After a great deal of effort by Dieterly and others [10.2] [10.3] an effective path length of 94 cm has been decided on. This is a middle of the road value. It will almost always be incorrect to some degree, but if it is used as a conventional figure workers in different laboratories can get the same results as each other. The errors arising from the use of a conventional path length have been found to be more tolerable than the prospect of having to make special allowances at every test.

10.4   Density

If the weight of a sample is being used in conjunction with width and length to calculate its cross-sectional area then a value of density must be used. It is notoriously difficult to determine density accurately so rather than attempt this for each test a system of conventional densities is used. Conventional values for density are agreed internationally and incorporated into product standards. Advanced electromagnetic methods of cross-sectional area determination exist and may well find increasing application.

It has been found that steel strip, if driven hard into saturation by some 70 kiloamps/metre of applied field, reaches a J_{sat} figure which is dependent on cross-sectional area and chemistry alone.

Experiment has shown that if 'conventional' J_{sat} values are allocated to steel types then the emf produced by flux reversal from high saturation values gives exact figures for cross-sectional area. This is just what is needed for setting the $\hat{B}$ values in an adjacent on-line power loss tester or in an Epstein test. The frequency of reversal can be 50 Hz for convenience and no waveform control is needed.

Most of the current demand of the system is reactive so that power factor correction reduces power demands to acceptable levels.

10.5   Overview of the Epstein test

Although the method sounds simple, poor results arise unless strict precautions are taken. These include:

- Accurately dimensioned square
- Accurately cut and dimensioned samples (annealed if required) so that mass and cross-section are accurate to ±0.2%
- Source of excitation and waveform control able to secure a sinusoidal form factor of 1.11 ± 1%
- Exact frequency control of excitation – better than ±0.1%
- Accurate $\hat{B}$ setting system
- Accurate wattmeter able to work at low power factors, e.g. PF < 0.1
- Precise airflux compensation (usually needed to be set once only when a frame is made)
- Firm grasp of conventional density, path length and calculation formulae.

When the material is thin, air gaps may open up at the corners due to 'sag' in the limb length, etc. Then small weights up to 100 g are sometimes applied to each overlap corner.

It should be mentioned that samples should be carefully demagnetised in a diminishing AC field before test to wipe out any previous magnetic history the sample may have.

A full exposition of the Epstein test system can be found in BSI, ASTM and IEC publications.

10.6   The single sheet tester

Electrical steels are a high-cost specialist product so their purchase and sale are conducted against strict quality standards. Normally power loss is the defining property to determine a material grade.

The rising volumes of material used and the associated testing means that the cost and convenience of testing for grading is an important matter. The Epstein test has been of long-standing value to design engineers who have learned how to translate the figures arising from the Epstein test into useful predictions of machine performance. A climate of change, however, has developed.

Shortcomings of the Epstein test are:

- Its path length is conventionally determined so that it does not necessarily equate well with the effective path length within the active volume of steel in a machine.
- Strips 3 cm wide involve cut edges only 3 cm apart so that unless the effects of cutting stress are accepted an annealing treatment must be applied before test.
- In many instances steel is used in the as-finally-processed state and an anneal at the final cut plate stage is not used. Further, some steel such as non-heat-proof domain controlled steels cease to show their proper properties if stress-relief annealed.

  In the case of grain-oriented steel, all limbs of the Epstein frame are loaded with strips cut along the rolling (easily magnetised) direction of the steel. For notionally non-oriented steel, opposite limbs are loaded with 'with' and 'against' rolling direction samples. Figure 10.5 shows this.

  Notionally non-oriented steel always has some anisotropy so that disparate permeability in adjoining limbs leads to extra flux leakage at corners.
- The preparation of samples using precision shearing and annealing is costly and labour intensive, as is the testing itself involving loading and unloading of test frames. Attempts to automate this process have met with limited success.

Many of the drawbacks of Epstein testing are overcome by use of the single sheet tester. This device operates on single sheet samples chosen to be of a size representative of production material properties and large enough to make effects

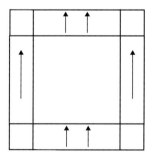

Figure 10.5   Loading of an Epstein frame with mixed 'with' and 'against' steel strips

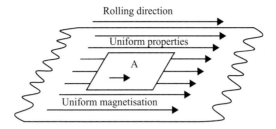

Figure 10.6   A is a notional area representative of strip properties

due to cut edges insignificant. A great deal of effort has been expended over the last 30 years towards developing a single sheet tester which satisfies the needs and aspirations of the commercial world.

10.6.1   Sample size

At the start of discussions and negotiations on this matter participating countries had in informal use a wide range of testers. This involved samples ranging from 1 metre wide × 1–2 metre long, to samples a few centimetres wide and perhaps 30–40 cm long. Figure 10.6 indicates the existence of a notional region within production strip whose properties are to be assessed. If this is too small it will be unrepresentative, if too large the costs of handling and test (see below on flux closure) will be excessive.

10.6.2   Magnetic circuit

A wide range of magnetic circuits have been in use, some with flux closure yokes and some without. Figure 10.7 sketches some of these. If flux closure yokes are absent it will be difficult to secure a region of uniform magnetisation within which tests can be made. If flux closure yokes are used these can be single-sided or double-sided. If single-sided, eddy current pools can form in the strip where flux leaves or enters it normal to its surface. These pools of eddy currents falsely add to measured losses. With two-sided yokes these eddy effects largely cancel (Figure 10.8).

If samples are large, e.g. 50 cm square, the yokes involved are very heavy and expensive. The top yoke of a tester can be raised and lowered pneumatically and counterbalanced so that the pressure on sample ends is not great. Having decided on a yoke configuration a working path length for the device must be decided upon. The conditions of fringing flux near the yokes will vary in shape and intensity as the steel permeability and thickness may vary, etc. Of course, a notional path length could be chosen and caused to apply to all samples using the same fit-all policy as applies to the 25 cm Epstein.

At this point a difficulty arises. Consumers of test data will be likely to compare figures from Epstein and single sheet methods. If a single conventional path length is chosen for a single sheet tester it will be mostly incorrect, and will not align exactly

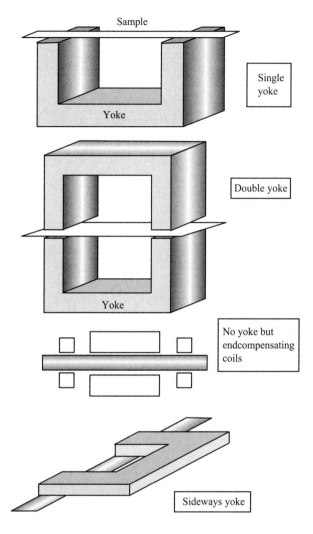

Figure 10.7   Flux closures

with the almost-always-wrong Epstein. This need not be a severe problem if the nature of a single sheet test were decided and brought into universal use while user engineers were open to re-learn the way in which the new (not-very-different) figures mapped into real machine performance. Further, sample sheets are most convenient if square, so that one can be inserted in the tester, then re-inserted after 90° rotation so that the residual anisotropy of notionally non-oriented steel can be taken into account.

With a square sample the low length/width ratio means that flux 'wander' along the active path length can become a problem, Figure 10.9.

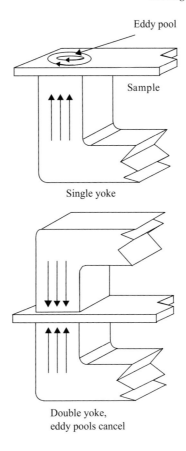

Figure 10.8   Eddy current pools due to normal flux

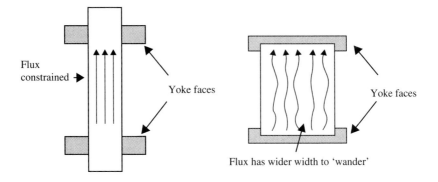

Figure 10.9   Flux wander in test sheet

Sheet enters here

Figure 10.10　Commercial single sheet tester (cover removed)

Testers able to measure simultaneously in the 'with' and 'against' directions have been devised but are complex to make and use.

Current international standards use either a conventional path length for single sheet tests or allow the applied path length to be varied with material type and grade in such a way that as close as possible agreement with the 25 cm Epstein test is obtained.

There has always been a tension between parties desirous of getting the most precise scientifically based measurement and those wishing any test to be not necessarily perfect, but inexpensive, convenient and capable of giving adequate engineering information so that machine design can be confidently carried out based on measurements used for purchase and sale.

The current situation is that single sheet testers are embodied in international standards having a sample size of 50 cm square and using double yokes. Path length is either taken to be the inside-of-yoke width, or calibrated to give figures aligning with the 25 cm Epstein frame.

10.6.3　Special considerations in single sheet testers

Figure 10.10 shows a commercial single sheet tester. It has primary and secondary windings, air flux compensation and double yokes.

10.7　Excitation

A great deal of effort has been expended on studies of how to best magnetise the sample in a single sheet tester. If a single winding along the whole length of the

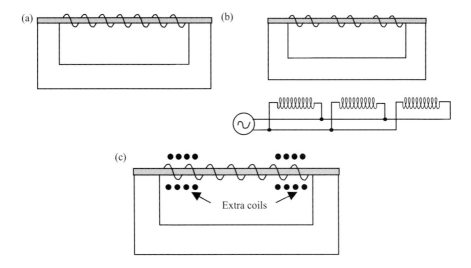

Figure 10.11 Use of parallel sections and end overwinds
   a Single winding
   b Multiple parallel windings
   c Extra overwind near ends

sample is used (Figure 10.11(a)), there is a tendency for attained flux levels to fall off near the ends, so that if magnetising current is used as a measure of H, and to feed a wattmeter, the demagnetising effects of the sample – yoke reluctance are not accounted for. One device (b) to minimise this effect consists of magnetising the strip by using a plurality of windings in parallel. The sections near the ends automatically take more current as they offer a lower reactance to the supply due to end-of-sample demagnetising effects.

Improved uniformity of magnetisation can be obtained by adding extra turns near the sample ends so that the extra mmf arising cancels the effect of joint reluctance (Figure 10.11c). The current supply to the extra turns may more than compensate for the end reluctance and lead to errors, so that special means may be used to get this compensation exactly correct. The end compensation windings can be separately supplied in response to the signals from a magnetic potentiometer (Rogowski–Chattock coil) suitably placed. Figure 10.12(a) outlines the system.

Further, uniformity of magnetisation can be improved by the use of Dannat plates [10.5]. These are copper sheets mounted inside the windings above and below the sample, see Figure 10.12(b). Any tendency for flux to diverge from the sample causes it to meet these Dannat plates normal to their surfaces and induce eddy currents in them, which by Lenz's law action powerfully oppose the flux concerned. Dannat plates thus aid uniformity of magnetisation by a simple passive device, Figure 10.12(c).

An alternative strategy is to avoid using the magnetising current as a measure of H applied and to use so-called H coils near the sample surface to sense the effective

(a)

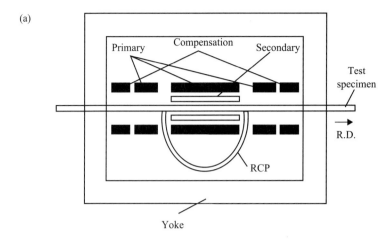

(b)

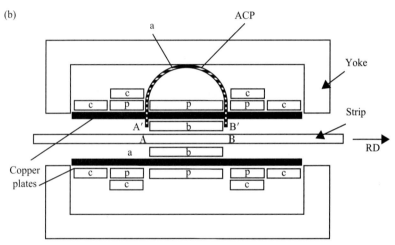

Cores-section through the test frame, showing position of primary (p),
secondary (b) and compensating (c) windings with respect to the yokes and strip

(c)

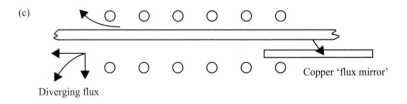

Figure 10.12   *a*   Closed yokes and Rogowski–Chattock potentiometer
   b   Open yokes and copper plates
   c   Restraint of diverging flux

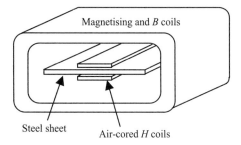

Figure 10.13   Location of H coils

applied field. This technique has the attraction of greater potential accuracy but is more complex to operate. Figure 10.13 shows the method in which an *H* coil is used. The output from the *H* coil is proportional to d*H*/d*t*, but to give a figure for loss, the quantity $\int H \, dB$ or $\int B \, dH$ is required.

B coils give an emf proportional to d*B*/d*t* and, combined with a signal related to *H* derived from the magnetising coil current, the $\int H \, dB$ is carried out in a conventional wattmeter of whatever sort. An *H* coil gives d*H*/d*t* so needs pre-integration before feeding a conventional wattmeter, otherwise the *B* coil signal d*B*/d*t* may be integrated to give a signal proportional to *B* and combined with d*H*/d*t* from an *H* coil to enable an integral *B* d*H* to be used.

There are many considerations relating to the phase accuracy of integrators, signal–noise ratios, mean/peak ratios of signals, slew rates of amplifiers, etc. which have to be managed so that a system is chosen giving reliability and accuracy. If *H* is assessed from a centrally placed *H* coil, Figures 10.14(a) and 14(b) , then a *B* signal derived from a *B* winding in the same region may allow concerns about end reluctances, etc. to be discounted as d*B* and d*H* signals are used from the same central area of uniformity.

It is accepted that to be most accurate an *H* coil has to be in intimate contact with the steel surface and very thin. A thin coil is comparatively insensitive. Placed flat on the steel surface an *H* coil is vulnerable to damage. The errors arising from being spaced away from the steel and protected by a strong thin non-magnetic cover can be mitigated if two *H* coils are used one above the other, see Figure 10.15. By noting the decrease of signal as between coil (a) and (b), a regression can be carried out enabling the signal that would have arisen from intimate contact to be computed.

So, some of the key features to be considered in single sheet testers are:

(i) Yokes or not? Two yokes or one only?
(ii) Magnetising windings single layer or sectionalised for ωL equalisation?
(iii) Compensation end windings to be used?
(iv) RCP Rogowski–Chattock coils to be used to adjust compensation?
(v) *H* coils used or not?
(vi) Double *H* coils considered?
(vii) Path length to be of a conventional value?

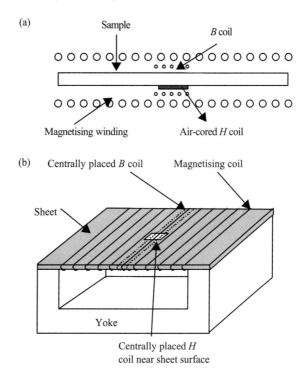

Figure 10.14   a   Placement of H coil
b   Sensing to avoid end effects

(viii)   Path length to be 'adjusted' so that figures closer to 25 cm Epstein results
        arise?
  (ix)   Sample to be how big?
   (x)   Dannat plates to be used? [10.5]

Current international standards favour:

- 50 cm square sample
- Single or sectional magnetising coils
- Two yokes
- *H* from magnetising current not *H* coils
- No Dannat plates.

It may be commented that yokes for 50 cm square samples are large, heavy and
expensive to make. The yoke faces have to be carefully prepared so that the laminations
of which they are made do not experience in-surface short-circuits. Yoke faces can
be ground flat and chemically etched to remove 'smeared' metal. Power loss in the
yokes is very small but can be allowed for.

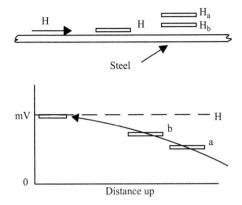

Figure 10.15   Double H coil placement

10.8   Air flux compensation

Just like the Epstein frame a single sheet tester requires air flux compensation. It can be done by a mutual inductor as for the Epstein but there is a complication. When the empty tester is excited and the net B coil plus air mutual secondary output is noted, adjustment of the inductor may lead to a balance that is dependent on the size of the current in the magnetising windings. This can arise if the adjacent iron yokes become slightly magnetised by the excitation and deliver consequential flux to the B coil.

An alternative method of adjustment can be carried out with a sample present but waveform control omitted. The progressive adjustment of the mutual inductor leads to waveforms from the B coil/mutual secondary as shown in Figure 10.16.

(b) is undercompensated; (c) is overcompensated as shown by an inversion near $dB/dt = 0$, i.e. $\hat{B}$; (a) is correct.

10.9   On-line testing

Even though single sheet testing is a lot more convenient than 25 cm Epstein testing there is a strong motivation to look forward to its being superseded by direct grading of material on the production line. The continuous measurement of magnetic properties on a steel strip line is a difficult task but it can be done and the technique is in routine use within the industry. On-line measurement poses the following basic problems:

(a) How to magnetise the strip
(b) How to sense its state of magnetisation
(c) How to be aware of the strip cross-sectional area so that correct working inductions can be set.

Various strategies have been used for the magnetisation of strip. Figure 10.17 shows two of them. Figure 10.17(a) uses a coil wound round the strip with or without

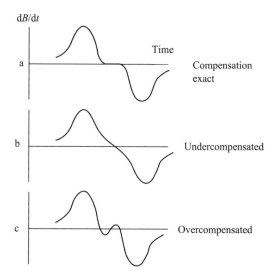

Figure 10.16   Adjustment of air flux compensation

flux closure yokes. Figure 10.17(b) shows wound yokes used to inject flux into the strip. On the principle that it is most effective to locate magnetomotive force next to the strip, (a) is to be preferred. Magnetisation can be sensed by use of a *B* coil (d*B*/d*t* coil) wound on the same former as the magnetising coil. Taking note of the factors affecting the design of single sheet testers, on-line testers could:

(a)   use multiple parallel operated magnetising coils
(b)   use a single magnetising coil only
(c)   use a single coil with extra end windings having excitation controlled by the findings of a RCP
(d)   use the current in the magnetising windings to provide an *H* signal
(e)   use *H* coils, single or double, to develop an *H* signal
(f)   yokes may be single, double or absent
(g)   Dannat plates may be used or not.

10.10   Enwrapping/non-enwrapping systems

The coils shown in Figure 10.18 enwrap the strip so that to remove the tester for calibration or maintenance involves cutting the strip. Considerable efforts have been devoted to devising testers which magnetise the strip and sense its induction without enwrapping it. Figures 10.18 and 10.19 outline the development of such devices. These techniques are considered in detail in [10.6] and [10.7].

A tester which does not enwrap the strip will also allow magnetisation to be carried out at 90° (or other angles) to the rolling direction of the strip. Measurement at 0° and

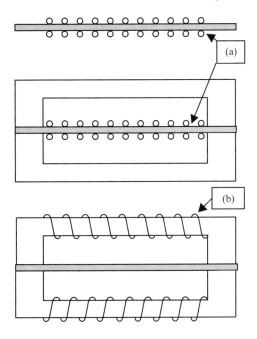

Figure 10.17   Methods of magnetisation

90° or at intermediate angles allows the residual anisotropy of notionally isotropic strip to be measured. Figure 10.19 shows the early development of non-enwrapping magnetisers.

Versions of on-line testers have been developed in which both magnetisation and measurement is undertaken from one side of the sheet. This means that on a production line the test device can be inset into a surface over which steel passes so that no obstruction to line operation can arise. The effects of an incomplete magnetic circuit and flux leakage is addressed by appropriate algorithms and neural net systems which sense magnetising current as an indication of permeability. Flux leakage is a function of permeability and appropriate adjustments can be made by computer.

10.11   Cross-sectional area of strip

To be able to set the required $\hat{B}$ value of strip in a production line the cross-sectional area of metal must be known so that the appropriate average voltage can be noted from the B sensing coils. The width of strip in production is normally exactly known, though a minute amount of stretch can occur if steel is under tension when hot. The thickness of strip is needed for other quality control reasons so that an accurate thickness meter is required on line.

The combination of known width and thickness from the thickness gauge enables a cross-section figure to be given to the tester $\hat{B}$ setting circuitry.

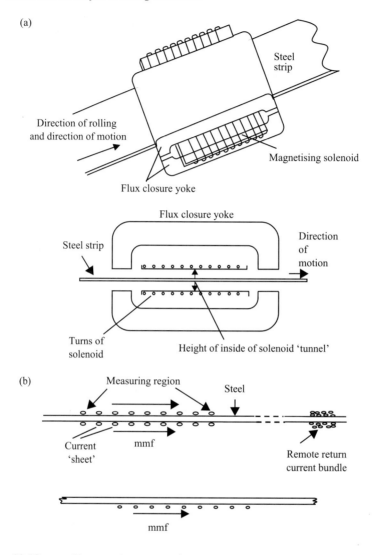

Figure 10.18 *a* Enwrapping magnetiser
           b Non-enwrapping magnetisers

In Epstein and single sheet testing 'conventional' densities are used with weighing of samples. The thickness figure from a thickness gauge involves the gauge being advised of the conventional density for the grade of strip in process. The whole question of thickness and conventional densities will be examined in Chapter 13. If magnetic means are used to assess cross-sectional area 'conventional magnetic constants' must be used (see Chapter 13).

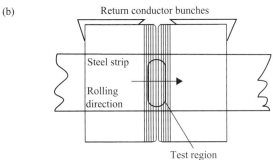

Symmetrical non-enwrapping magnetising system.
A similar set of conductors is placed below the strip.

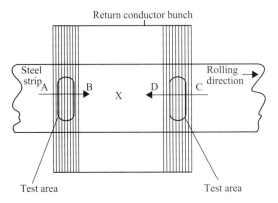

Figure 10.19   *a*   Development model of a non-enwrapping magnetiser
                b   Non-enwrapping magnetisers

10.12   The earth's field

Long strip in a production line is exposed to the earth's magnetic field which can give a DC magnetic bias to measurements. It is not always convenient to build production lines in an east–west direction, so that it becomes convenient to add a countervailing DC bias to the AC magnetisation field used for measurement. Electronic checks to verify the symmetry of dB/dT waveforms can be incorporated into automatic earth's field correction systems.

The origins of the earth's field developed from the 'dynamo' at the earth's centre which involves its liquid core remain obscure, particularly as reversal of the earth's field has occurred many times during pre-history.

10.13   Challenges

The challenges facing the measurement and grading world are these:

(i)   The Epstein test is expensive in cutting, annealing, handling, etc. and its path length is arbitrary. Engineers are used to allow for the effects of inappropriate path lengths but better practice should be possible.

(ii)   Single sheet testers offer a test closer to service conditions. Problems of assigning a path length to single sheet testers remain, especially if it is hoped to coax the device to yield Epstein-like results. A range of design possibilities exist, single or multiple H coils, RCPs, etc. Simplicity is attractive but since computing power is cheap and available more ambitious systems could be used involving various error-reducing techniques.

(iii)   It is inevitable that eventually grading will be done by on-line continuous assessment; economics will drive the change. If the design of such testers has as much potential for controversy as the single sheet tester the outlook for convenient universality of test is bleak.

Currently the single sheet test is becoming more used but tends to be looking backwards towards alignment with its Epstein predecessor. It would be well to further develop the single sheet test with foresight as to the likely nature of on-line testers. A realistic aim for the first decade of the current millennium would be a test scene in which single sheet tests aligned well with on-line devices because their mutual design had been conceived that way. Thereafter the Epstein test acting as a laboratory tool for special research could possibly acquire H coils itself.

The separate challenges of reconciliation between on-line tests and the grading of semi-finished non-oriented steels can be tackled in various ways. Data collection is now sufficiently refined to make predictive algorithms possible for gauging the quality of not-finally-annealed steel from magnetic measurement carried out before critical reduction and a measurement of the amount of reduction used.

10.14   Other test devices

10.14.1   Ring tests

The use of a totally closed magnetic circuit is attractive for test purposes: no lap joints, etc. are involved. In many ways the ring test is the most fundamental though perhaps the least convenient. Given a *very* long solenoid, Figure 10.20, the region XY could be considered remote from the ends and tests carried out there. More practically the metal can be formed into a ring, Figure 10.21. The ring can have various forms such as:

(a)   stack of stamped-out rings
(b)   coil spring format
(c)   welded joint
(d)   'flattened ring' to enable Epstein samples to be cut from it (Figure 10.22).

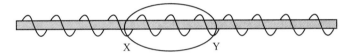

X   Y

This area free from end effects

Figure 10.20   'Long sample'

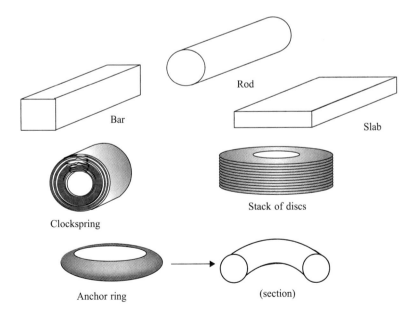

Bar

Rod

Slab

Clockspring

Stack of discs

Anchor ring   (section)

Figure 10.21   Sample forms

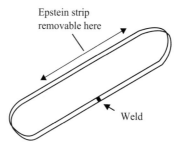

Figure 10.22   Flattened ring able to be wound and tested toroidally, and have an Epstein strip cut from the long side for comparative tests

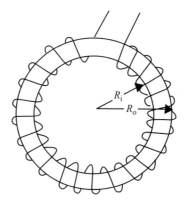

Figure 10.23   Ratio of diameters in a ring ($R_i = R_{inner}$, $R_o = R_{outer}$)

In all these cases annealing must be applied to remove the stress of cutting, bending, welding, etc. In the case of (a) the stack can include rings placed above each other in such a way that any anisotropy of metal can be averaged out round the ring. Unless the ring has an outside/inside diameter ratio of less than 1.1 : 1, the path length will vary with diameter such as to cloud the measurement results (Figure 10.23).

Winding coils onto rings can be done by use of a toroidal winding machine (there may be a need for a 'box' round the metal to avoid stress due to applying the winding).

Consider also:

- use of a multi-pin fitting (Figure 10.24);
- use of a multi-strand 'rope' of wires able to be wound on, plugged together at the ends for ease of application and removal (Figure 10.25).

Once the ring is made and wound, measurements can follow the usual procedure with a wattmeter, etc. Air flux compensation is not needed if a *B* winding is closely applied to the ring.

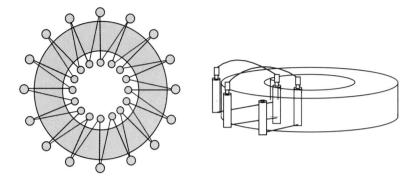

Figure 10.24   Interlocking pins combine to give a 'winding' for a ring sample

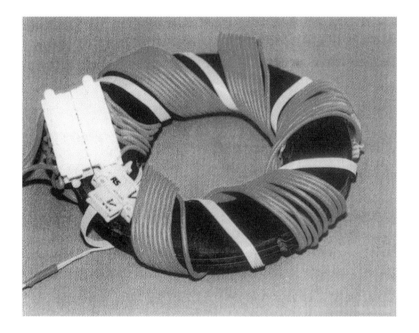

Figure 10.25   Removeable 'rope' winding, rapidly applied and ends connected by a plug and socket

Approximate but very rapid measurements can be made using a 'single-turn' magnetisation device in which a heavy current is passed down a copper contact set. A *B* sensing wire can be included in the fitment. In this way whole cores can be tested rapidly on a production basis, Figure 10.26. The device is made so that leakage of magnetising current into the *B* sensing circuit is negligibly small.

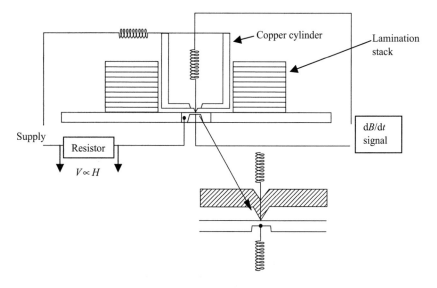

Figure 10.26   'Copper plunger' method for stack measurements

10.14.2   Portable testers

Over many years a range of portable testers has emerged which can take a rapid measurement of core loss when 'placed-on' a sheet of steel. An early one, the Werner tester, used the damping of an oscillating circuit produced by the steel to give an indication of loss. Figure 10.27 shows a range of other testers. All these devices have their usefulness but cannot command the accuracy of larger-scale devices.

There are various methods of measurement of the watts dissipated in a test system. The principal ones are:

(a)   the dynamometer wattmeter
(b)   the stochastic–ergodic method
(c)   the electronic multiplier
(d)   the bridge method
(e)   the fully digital method.

The dynamometer wattmeter is the most traceably fundamental and involves the forces acting between two conductor systems, one carrying current proportional to H and the other to dB/dt. Typically this method will be realised in the form of a suspension carrying one coil able to rotate, set within a set of static coils. The suspension is very delicate and uses a mirror-galvanometer system to observe deflections.

No iron may be used for flux magnification and the reactance of the windings must be kept to levels which do not lead to serious phase errors. A good wattmeter should be able to deliver 0.5% accuracy down to power factors of 0.1 or less.

The upper frequency at which such wattmeters can operate tends to be limited to a few hundred hertz. However the device responds down to zero frequency so

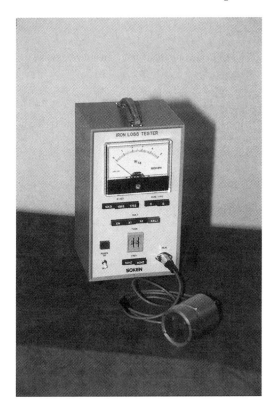

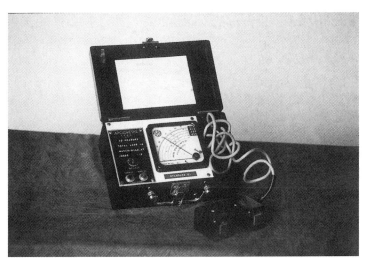

Figure 10.27   Portable loss testers usable on small samples via a 'place-on' yoke

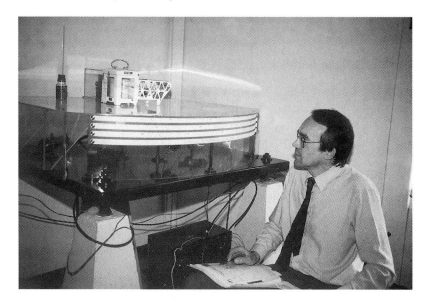

Figure 10.28   Standard dynamometer wattmeter

that calibration can be linked to standard cells and standard resistors. This facilitates traceability of calibration. Figure 10.28 shows such a wattmeter. A good account of wattmeters is given in Ref. [10.8].

10.14.3   The stochastic–ergodic system

This relies on the properties of interacting noise signals to give a multiplicative effect. The theory of operation is complex and is given in Ref. [10.4]. Such devices are flexible and convenient in operation.

10.14.4   The electronic multiplier

This uses electronic operational units to deliver a product of input analogue signals. Great care is needed to minimise the effects of drift and offsets.

10.14.5   Bridge methods

These depend on the fact that a steel sample enwrapped by windings has an equivalent circuit of the form shown in Figure 10.29. The resistance relating to iron loss can be measured if the impedance of the whole device is determined by balancing a bridge network. Separate balances are required for the reactive and resistive elements of the sample and coils.

In former times a large technology of bridge devising and management grew up and a range of variations came into use. All had various advantages

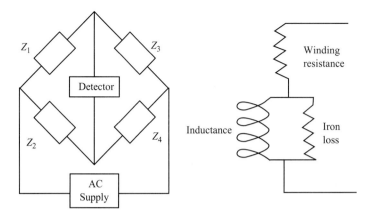

Figure 10.29 *At balance the detector reads zero when $Z_1 = Z_2 \times Z_3/Z_4$. By variation of Z_2, Z_3 and Z_4 a balance may be secured and Z_1 evaluated*

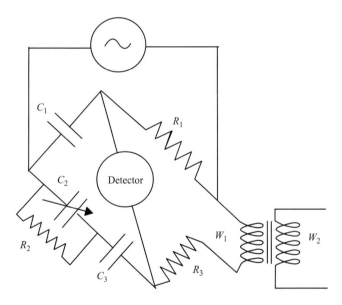

Figure 10.30 *Bridge for loss measurement. At balance, C_2 is balancing the inductance of the square winding and R_2 balances the iron losses. Winding W_2 can be used for monitoring the operating induction of the sample, or for the application of DC bias to the sample. C_3 balances the effect of R_3, the square winding resistance*

and disadvantages. Figure 10.30 shows a typical bridge. By appropriate configuration of windings an Epstein frame could form a bridge component and allow the power loss of the steel up to frequencies of 10 kHz or above to be determined.

Today bridges are out of fashion due to their complexity and need for skilled operation. Computer methods (expert systems) may find them in use again in the future. Ref. [10.3] gives useful guidance as do the various editions of *Alternating current bridge methods* by B. Hague.

When fully digital systems are used for magnetisation and induction setting, the determination of loss can be done directly from the captured data available during each cycle of operation. As with many modern systems they are very user friendly when working properly, but if faulty can be very hard to unravel faults.

From time to time thermal methods have been used to assess power loss. This could include a 'constant flow' method in which the rise in temperature of a coolant is used to measure the heat evolved, or the rate-of-temperature-rise method in which the rate of rise of emf from a thermocouple (or array of thermocouples) attached to a sample can be noted. The initial rate of rise of thermal emf is little affected by thermal leakage as the response is adiabatic. The first few seconds of excitation are used for measurement [10.9]. Early in the century Ewing devised a hysteresis tester which was hand-operated and delivered an immediate read out. Ref. [10.10] describes its operation. A steel sample was spun by a hand-driven rotator between the poles of a pivoted permanent magnet. Hysteresis effects caused the magnet to deflect and move a pointer over a scale.

It is to be hoped that oncoming methods will in due course be as easy to operate, yet as precise as manufacture and commerce may require. It would help enormously if any overarching standardisation body were able to authoritatively map the way forward on commercial grading so that duplication of effort and clashes of interest could be avoided.

There is a growing use of 'virtual' instruments for all sorts of measurement, B, H, watts, etc. which are embodied in appropriate 'cards' and a PC. A critical evaluation of the real achievement specification of such 'instruments' is required to ensure that limitations of dV/dt, slew rates, dynamic range, etc. do not lead to unsuspected errors.

Key components of testing strategy could include:

- relegation of the Epstein test to laboratory and research duty
- evolution of a single sheet test decoupled from the Epstein and allowing easy extension to compatibility with on-line systems
- an on-line continuous testing method designed to be compatible with a developed single sheet tester. A single sided non-enwrapping version would be most convenient.

Design engineers would rapidly assimilate revised grade numbers so that prediction of machine characteristics would be satisfactory. As always questions arise over where resources to bring these developments to fruition may be found. At least computing power is cheap and becoming more so, so that some complexity could well be tolerated.

10.15   Rotational loss

Many parts of machine cores, such as transformer joints and the stator tooth roots of motors, carry elements of rotational loss. The *B* vector has a rotational as well as a linear component. The extra loss arising due to rotational loss has been widely studied, but especially in anisotropic metal like grain-oriented steel, the spatial variation in permeability makes it difficult to sustain and control a rotating flux vector of constant defined amplitude and waveform. Certainly in actual devices, rotational flux is always associated with waveform distortion and amplitude variations. It may be expected that steels which perform equally under linear magnetisation may behave differently when exposed to rotational flux.

Much more work will be needed before characterisation of materials in terms of rotational loss may become convenient. The views taken of rotational loss vary widely. One school of thought considers that knowledge of the linear losses (longitudinal and transverse) is sufficient for designers to adequately estimate behaviour in a given machine design. An alternative view is that it is important to be able to actually measure rotational losses. This can indeed be done and an extensive research literature exists on the topic but problems abound. For instance, is a pure constant amplitude *B* vector to be created and rotated? This can be imposed by computer methods and powerful drive amplifiers, but in real machines such a regime never identically applies.

Losses are noted to differ if vector rotation is clockwise or anticlockwise, due probably to unevenness in grain texture. The directions of applied magnetising field and attained magnetisation vectors do not coincide and angular separation varies as the *B* vector moves. The existence of multivalued *B–H* relationships complicates the interpretation of results.

So far knowledge of the directional permeability curves and loss versus $\hat{B}$ plots cannot be fed into FEM software in such a way as to usefully predict device losses, particularly when spatial discontinuities in material at joints occur and air gaps are present.

While there is a rising tide of demand for standards and standardised test methods for high-frequency pulse excitation of steel (e.g. for PWM applications), there is only a weak demand for rotational testing. This situation remains open and time will show how the balance of research interest and hard industrial application need turns out.

10.16   Measurements other than power loss

10.16.1   *Specific apparent power – volt amps per kg – VAs*

Designers of machines are interested not only in the power loss arising when steel is cycled through its peak operating induction, but in the amount of magnetomotive force required to do that. To take an extreme example, use of air and no iron at all would eliminate iron loss but require enormous currents to create useful flux levels. So the permeability of the steel matters as well as its power (core) loss. If two steels of equal power loss at the same frequency and peak induction require widely differing magnetising currents, then the losses happening in the copper windings used will differ widely also.

Power dissipated in a winding varies as I^2R and if more current is demanded this loss rises. A useful measure of this penalty is the so-called specific apparent power or volt–ampère product. If steel in an Epstein square or other test device is excited to a given peak induction then the product of the voltage associated with the number of turns and the RMS (root mean square) current required will be indicative of the $V \times A$ product or apparent power demanded.

For any electrical device the watts consumed $= V \times I \times$ power factor. If the power factor is low, the number of real heat dissipating watts is low but the current flowing in the winding can be much higher than provision of that number of watts would imply. $V \times A$ is the volt-ampere product and for a set supply voltage shows what the current drawn will be, and via the circuit ohmic resistance, what the I^2R losses will be.

When a sinusoidal induction is enforced by the use of a feedback amplifier the RMS value of the voltage at the square secondary if multiplied by the RMS value of supplied current gives an accurate value for the *VA* product.

Note: Average rectified voltage is noted for $\hat{B}$ setting, but RMS for *VA* calculation.

The RMS value of current supplied may be noted from an RMS-sensitive ammeter suitably connected. If the number of turns on the secondary winding differs from that on the primary, suitable arithmetic adjustments must be made to the $V \times A$ product. Then when $V \times A$ is divided by sample mass, a VA/kg value for specific apparent power emerges.

10.16.2 Permeability

While the VA product involves the permeability of the steel the specific quotient B/H is of interest for some applications. For thin material at power frequencies a plot of $\hat{B}$ versus $\hat{H}$ yields a curve not dissimilar to the traditional normal induction curve and represents a rapid method of getting a view of what a normal induction curve would be like. Particularly above 1.0 T the similarity is good.

The H peak can be determined by the use of:

• average rectified voltage sensing of an H coil output (may need amplification)
• use of a peak reading ammeter in the magnetising current feed line. This could be a peak reading voltmeter in parallel with a precision resistor.

Plots of power loss versus $\hat{B}$ and of power loss versus frequency give an excellent insight into the properties of an electrical steel. Further, plots of B/H and of *VA* versus $\hat{B}$ complete the picture.

10.16.3 Direct current measurements

The tests so far described are all related to AC conditions. Electrical steels are used in DC applications also. These may include relays, nuclear particle accelerator magnets and the like. For these applications power loss is not a prime consideration, but other parameters are important. These may be:

- coercive force
- saturation induction
- remanent magnetism
- normal induction curve shape
- maximum and initial permeabilities.

These are set out in Figure 10.31.

The relevance of coercive force may be illustrated as follows. Small alternators used in portable generators and the like are expected to self-excite at start-up. At start-up the rotor winding is fed with DC derived from rectification of stator output which is present due to the remanent magnetism in the machine flux. To work well a high remanent magnetism (associated with a high coercive force) is needed. Stator steel needs to have low power losses, so low hysteresis, low coercive force metal is used. However the rotor may need to be made of a higher coercive force material so that excitation will occur. The need is greater in more recent years where rectifiers have changed from germanium to silicon, which has a higher forward breakover voltage.

The design features and metallurgy which allows (economically) the stator and rotor laminations to be derived from the same feedstock are covered by commercial confidentiality.

A range of techniques is available to measure these properties. The ring method and Epstein frame may be used to apply DC tests to samples as well as specialist permeameters. A permeameter is an arrangement of sample, coils and yokes specially adapted for DC tests. There are many types and varieties of permeameter. Figure 10.32

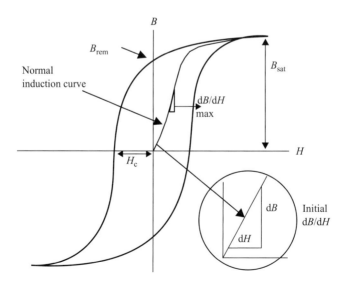

Figure 10.31   Key DC properties of an electrical steel evident from a BH loop

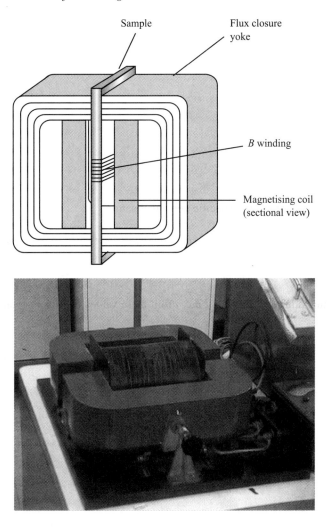

Figure 10.32   Diagram and photograph of a permeameter

shows one. Plate samples are placed in the coil system and yokes closed up to complete the magnetic circuit. The coils involve:

(a)   main magnetising coils
(b)   compensating windings to apply extra mmf to offset leakage at the yoke–sample joint
(c)   *B* coils wound onto the sample itself, or on a separate former
(d)   *H* coils if these are used
(e)   RCP system to assess exactness of compensation applied.

Magnetising fields can be applied via the main magnetising coils and if need be compensation via the compensating coils. A range of values of applied field can be used depending on the measurements required.

(a) A sample is placed in the coils and the yokes closed up.
(b) A strong magnetising field is applied to take the sample close to saturation. This is then reversed in polarity several times by a manual or automated switch to set the material into a reproducible cyclic magnetic state. This process is repeated for progressively decreasing values of applied field until the sample may be considered to be demagnetised. Alternatively a slowly decreasing AC field could be used for demagnetisation. However the high inductance of the windings of some permeameters may make this difficult. A freshly annealed sample may not require specific demagnetisation.
(c) A low value of applied field is selected, e.g. 40 A/m and a cyclic state established first by repeated reversals. Then the total flux change arising from reversing this field H_1 is noted by the charge circulating in a charge integrator connected to the B coils of the permeameter. This corresponds to a magnetisation change of $2B_1$.
(d) This procedure is repeated for higher values of H: H_2, H_3, etc. giving corresponding values of $2B_2$, $2B_3$, etc. From these values the tips of loops are used to plot out a normal induction curve.

At each point a check may be made to see that flux leakage compensation is exact if a facility is provided for using compensating windings.

The charge integrator used to determine B values can be a ballistic galvanometer or an electronic charge integrator. This will be calibrated such that with a knowledge of the sample cross-sectional area the number of B coil turns and the size of deflection produced, a value for the size of B reversal can be determined.

When samples are thick, e.g. 3+ mm, operations must be carried out slowly enough to allow full flux penetration into the sample (delay due to Lenz law effects). The permeameter might be configured so that some of the procedure is automated. Ballistic galvanometers and charge integrators can be calibrated by use of charge delivered from standard capacitors or mutual inductors [10.8].

10.17 The full hysteresis loop

This is explored along similar lines. The maximum applied field is used first to determine $\hat{B}$ at H_{max}. Then, by interrupting the current used, the flux change from $\hat{B}$ to B_{Rem} is noted. This procedure can be repeated for intermediate excursions from H_{max} to lower H values. This gives the flux change for each interval. Similar proceedings allow the whole loop to be mapped. If compensation is being used it may need to be reversed in polarity for some of the field 'throws'.

Alternatively the sample may be taken round the whole loop gradually by control of H while the charge integrator progressively notes the change in B. With an appropriate set-up, an XY plotter can be used to plot out the hysteresis loop. This device is then called a hysteresigraph. Today hysteresigraphs are more used than traditional switching permeameters whose operation is somewhat of a skilled art form. Unless in daily practice, it is fatally easy to forget which operation to do next.

10.17.1 Cardinal points

The really key points which describe a steel are:

- Saturation induction
- Remanence
- Coercive force.

Saturation induction is relatively easy to determine by applying a single field reversal and noting the flux change $(2 \times B)$. The field must be sufficient to technically saturate the test specimen.

Coercive force may be determined either from a permeameter or a separate coercimeter (see later), and since $B = 0$ at H_c, the value is not dependent on the leakage, or absence of it, of the magnetic circuit.

Remanence, however, is dependent on the magnetic circuit conditions. A ring is near perfect, but air gaps in Epstein frames, etc. produce demagnetising effects which cause the observed B_{rem} (remanent) magnetisation to vary considerably. This reflects the fact that the remanence values which will occur in real physical devices can be highly variable.

Given that H_{max} and H_c are easily determined (see H_c later in this chapter) and B_{rem} is uncertain, most of the useful insight into a material could be obtained by a plot of H_{max} and H_c, Figure 10.33, onto a skeleton BH loop and putting in an estimate for remanence.

It is unusual to learn much more about the material from performance of a complete BH loop unless the permeameter is compensated to allow for magnetic leakage. Designers often call for a complete DC hysteresis loop which is very costly and

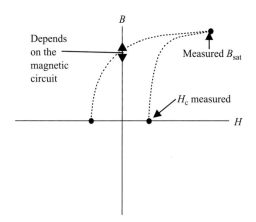

Figure 10.33 Estimate of magnetic properties

laborious to produce when knowledge of $\hat{B}$, H_c and perhaps an AC derived normal induction curve would suffice.

10.17.2   Coercive force

While the coercive force H_c can be determined from a permeameter a much more rapid and convenient system is available. This is the vibrating coil magnetometer. Figure 10.34 shows a schematic diagram of its set-up. The sample is mounted inside a long magnetising coil and placed so that one end of it is close to a small coil mounted on a piston arm driven by a vibrator. Whenever flux emerges from the end of the sample some of its diverging lines will be cut by the vibrating coil in which an alternating emf is generated at the frequency of vibration. When the magnetisation in the sample falls to zero the emf will fall to zero also. This happens at the coercive point where $B = 0$. The procedure of measurement is as follows:

(a)   Sample is put in place.
(b)   Vibrator is set in motion.
(c)   A strong current is applied to the system solenoid, taking the sample to a high value of B.
(d)   The coil current is then reduced progressively and caused to pass through zero and build up slowly in the opposite direction. When the output from the vibrating coil reaches zero the progress of current change is stopped and the value of current for B_0 read off. This current corresponds to an applied field which represents the coercive point for the sample. The magnetising solenoid can be calibrated so that current to the H value transfer and read out is simple and automatic.

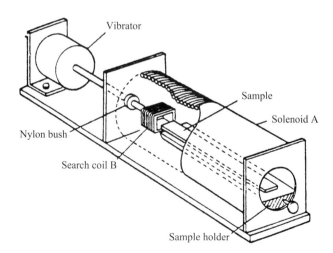

Figure 10.34   Mechanical arrangement of a vibrating coil magnetometer

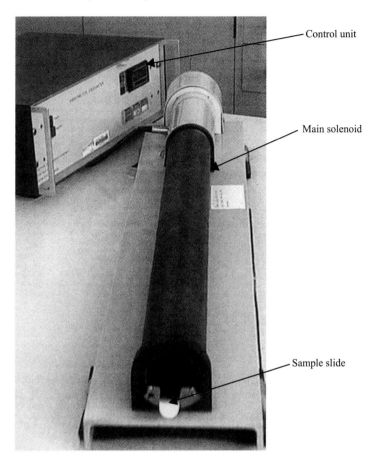

Figure 10.35   Vibrating coil magnetometer

(e)   The whole procedure should be repeated with magnetisation of the opposite polarity so that effects of the earth's magnetic field are eliminated when the two readings are averaged. By placing the device east–west the effect of the earth's field is minimised.

Commercial forms of this device are available which are automatic in action so that readings are rapidly obtained and read out. Figure 10.35 shows a typical device.

10.18   Mixed AC and DC excitation

So far consideration has been given to either AC or DC magnetisation. Sometimes a mixed situation arises where specific AC properties are required against a background of standing DC magnetisation. This arises in smoothing chokes, audio amplifiers,

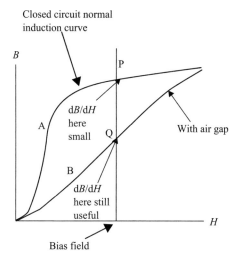

Figure 10.36   Normal induction curves with and without shear

magnetic amplifiers and related devices. Such conditions apply also to the pole faces of synchronous and DC machines, and in some varieties of flux switching motors.

Consider the normal induction curves in Figure 10.36. A is for a closed 'good' magnetic circuit such as may be obtained with fully interleaved laminations. B is for the same core but with an air gap. To attain the same induction at every point requires an extra mmf to overcome the reluctance of the air gap.

If the DC current took the core as in A to position P, the incremental permeability as indicated by a small minor loop would be low. In the case of B the same DC current takes the core only to Q. Here the incremental permeability is now quite a lot greater than at P. It is true that the absolute permeability B/H for the core has been reduced by the air gap, but the incremental permeability has been considerably enhanced.

In a smoothing choke or audio transformer where DC must flow, the incremental permeability dB/dH is the quantity that gives a useful value of inductance (henries) allowing smoothing or transformer action to occur.

Of course, by use of a much larger core without a gap an equivalent dB/dH could be obtained, but for economy of copper wire and core steel the use of an air gap is a valuable device for getting the required inductance as cheaply as possible.

The question arises of how to design the core gap and size of core. For a given grade of steel the quantities involved are:

L = required inductance
I = DC current to be carried
N = number of turns on the winding
l = magnetic path length of iron
a = air gap length
V = volume of the core.

The alternating component of current may be only a few per cent of the DC current or a sizeable proportion of it. The unique work carried out by Hanna [10.11] was to develop a set of curves into which the six variables above could be entered and this enabled an optimum gap length to be read off. A re-determination of such curves for the latest grades of modern steel combined with a computer program would make design easy. In the author's case cut and try experiments with paper taken from various newspapers to form a gap are well remembered.

10.19   Cores

Where electrical steel is wound into cores of various sizes whole cores can be evaluated by the application of temporary windings of the 'rope' variety. To avoid assembling a core for test, separate 'logs' (that is stacks of core plate making up one limb) can be tested singly. Work is currently going on to make such tests quick and convenient. A method for overcoming the severe effects of demagnetisation at the ends of logs is being actively pursued.

10.20   High-strength steel

Steel used in certain parts of the structure of large rotating machines needs to have high strength as well as a useful level of permeability. Such steels are often thick (1–3 mm) and are best evaluated in a permeameter or formed into a ring.

10.21   Amorphous metal

Amorphous metal has a very low power loss and high permeability. Being very thin it is inconvenient to cut it into many strips to fill an Epstein frame. On the other hand, wound cores tend to be stressed during the winding. Further, the combination of low loss and high permeability means that the power factor of excitation is very low and poses problems of stability in analogue feedback systems used for waveform control. Digital systems are reputed to be more tractable.

There is still a lot of work to be done before there is complete consensus on the best methods of test for amorphous material.

10.22   Traceability

It is important that the uncertainties in magnetic test and measurement should be accountable. The former British Calibration Service (BCS), now NAMAS, provides such a service. European Electrical Steels incorporates (at its Newport site) a NAMAS laboratory accredited for measurement of power loss, permeability, etc. Tests can be carried out for persons requiring them.

10.23 Availability of equipment

The setting up and operation of precision magnetic measuring apparatus using purchasable sub-components is a demanding task. European Electrical Steels manufactures and sells complete measuring equipment ranging from the basic to the highly automated. Much of this equipment is in use around the world. It is usual to discuss a user's needs and to produce a system most suitable for their requirements.

10.24 Surface insulation

The sheets making up the steel cores of transformers, motors and generators must be insulated from each other to restrain the flow of eddy currents. The interlaminar insulation is achieved by various means. For transformer steel a combined silicate glass-phosphate inorganic coating is used. It performs other duties besides insulation.

Many large machines employ a fully organic coating which cannot be exposed to annealing temperatures. Laminations which do have to be annealed can carry a mixed organic/inorganic coating. Small machines may use no formal coating, or only natural oxide or steam-induced blueing.

10.25 Test methods

There is a demand for quantitative information about the degree of interlaminar insulation afforded by various coatings. Consequently a wide range of test methods have evolved. Some are included in international standards.

A quite separate problem exists in setting levels of insulation deemed to be adequate for particular duties. Because exact models of the translation of interlaminar leakage currents into machine losses are lacking there is a tendency to specify higher and higher levels of insulation so as to 'be safe'. The provision of coatings represents a considerable expense so that there is a strong motivation to understand the matter as fully as possible.

The desire for a rapid result has often led to ad hoc tests being made, such as applying the probes of a commercial multimeter (ohms range) to a steel sheet surface (Figure 10.37). This is a quite uncontrolled procedure, the probe area and pressure are undetermined and the operating voltage hugely unrepresentative of the service conditions of the steel.

Attempts to formalise testing somewhat have led to the hand-held gripper (Figure 10.38). Here the area of pressure and the force applied are better controlled but the applied emf is likely to be some 1.5 volts – still much too high.

Recognising that steel is used in aggregated stacks, tests have been devised in which stacks of sheets are measured from top to bottom (Figure 10.39) sometimes with copper interleaves, sometimes without. Precautions to avoid layer-to-layer burr contact may or may not be taken and various applied voltages have been used. However present-day thinking and practice make use of three main devices.

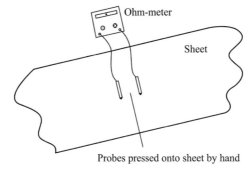

Ohm-meter and probes

Figure 10.37   Ad hoc insulation 'test'

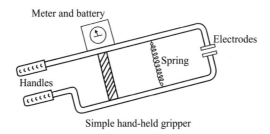

Simple hand-held gripper

Figure 10.38   Gripper test

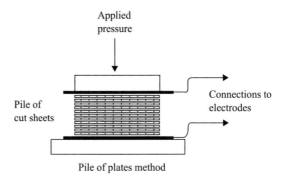

Pile of plates method

Figure 10.39   Interleave technique

10.25.1   The British Standard test

This was featured in the now obsolete BS 601 and is fully described in BS 6404 pt.20 1996. Figure 10.40 outlines the system. Contact probes of 645 mm^2 area were pressed together with a force of 450 N. A standard AC voltage was used for excitation, the

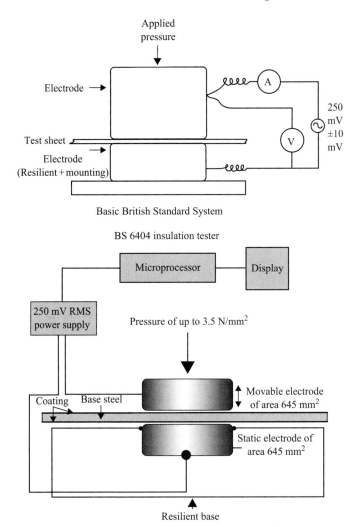

Figure 10.40   Details of the BSI type test

value varying over the years, 250 mV being the most recent. The current drawn gave an indication of insulation quality. Clearly two insulated surfaces are under test in series, though a subsidiary contact applied to the substrate would give access to one surface only for test.

To get a useful assessment of surface insulation a statistically significant number of readings must be taken, spread over the sheet being examined. There is a severe temptation to make too few tests and fail to comply with the statistical treatment set out in the relevant standard. In response to this, multi-electrode tester systems

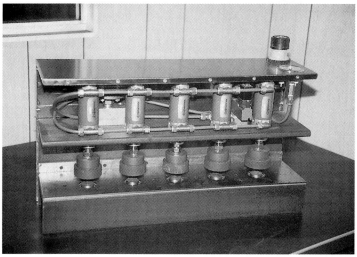

*Figure 10.41   Pneumatically operated BSI type insulation tester able to interrogate
five areas at one time*

have been devised by European Electrical Steels so that many readings may be taken
rapidly.

Operation and pressurisation of the electrode is done pneumatically. Figure 10.41
shows single and multiple electrode pair testers.

10.25.2 The Franklin test

In contrast to the British method, the Franklin test, originating in the USA [10.12], used ten button electrodes pressed on to one side of a sheet of steel, each electrode having an area of 64.5 mm^2 and pressed with a force of 129 N.

Each electrode was fed via a 5 ohm resistor from a 0.5 V source of DC. Figure 10.42(a) outlines the method. While the Franklin test was widely used 0.5 volts is a very high test voltage and depending on the surface insulation and current drawn, the emf at the electrodes will vary from zero to 0.5 volts. For perfect insulation the aggregated current drawn by the ten electrodes is nil, and for total short-circuit it is 1.0 amp. A twist drill cuts into the substrate and forms the return current path so that only one surface is evaluated at a time.

Obviously one electrode short-circuiting to substrate will swamp the fact that nine others may have perfect insulation. Again stringent use of statistical methods of operation is called for in standards before reporting a result. Human nature tempts the user to make one measurement.

A later development of the Franklin test currently incorporated into IEC standards alters the test conditions so that 250 mV (stabilised) is maintained at each of the ten buttons and the current drawn by each separately examined. Computer control of such a system in which each button is energised and monitored and its result recorded is included in an automated version of this device made and sold by European Electrical Steels (Newport).

Figure 10.42(b) shows one variety of Franklin tester for factory use. Analysis of the source of interlaminar emf shows that this is a constant (stabilised) amount linked to the induction attained in the steel, so that a low stabilised emf is appropriate for incorporation in an insulation test system. For an example of interlaminar emf see Figure 10.43.

Notional cross-section of region generating emf: $t \times w$. From

$$\tilde{V} = 4.44 \hat{B} n f A$$

$$\tilde{V} = 4.44 \hat{B} n f t w$$

for $\hat{B} = 1.5\,\text{T}, n = 1, f = 50, t = 0.3\,\text{mm}, w = 1\,\text{m}$,

$$\tilde{V} = 4.44 \times 1.5 \times 50 \times 0.3 \times 10^{-3}$$

$$\tilde{V} \simeq 100\,\text{mV}$$

The size of the worst case electric stress varies with the position and number of other leakages and short-circuits. Papers currently in preparation will address this matter in detail.

10.25.3 The Schmidt tester

It has long been recognised that the real situation in a machine involves two insulated surfaces pressed against each other rather than a single insulated surface in contact

(a)

Franklin insulation tester

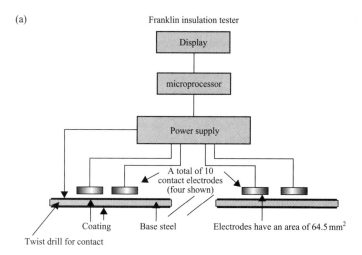

Twist drill for contact

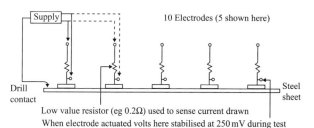

Low value resistor (eg 0.2Ω) used to sense current drawn
When electrode actuated volts here stabilised at 250 mV during test

Control computer excites each electrode in turn and records
current drawn for stabilised electrode voltage

Improvements to Franklin test

Figure 10.42   Continued

(b)

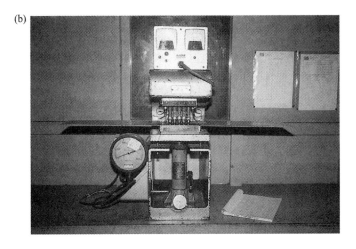

Figure 10.42 *a* Details of the Franklin test circuitry and electrodes along with twist drill
b An every-day factory model of the Franklin tester

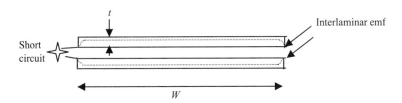

Figure 10.43 *It is convenient to consider emf arising from the metal enclosed by the dotted line, assuming a short-circuit on one side*

with a test electrode. With this in mind Dr Schmidt devised a test in which two sheets of steel are placed one on top of each other and separated by a paper mask, Figure 10.44, so that any burrs are kept away from each other. Contact is made to each substrate via clamp electrodes. A stabilised AC or DC emf may be applied to the two sheets via the clamp electrodes.

The assembly as described is now interrogated by applying a pair of insulated pressors in a pattern of positions within the mask area. This procedure simulates the circumstances of steel in an actual machine. The applied emf, shape and area of pressors and the treatment of data produced can be varied widely.

Because two coated surfaces appear 'in series' the measured insulation figures are a lot higher than for tests in which a single surface is interrogated. This is so, not only due to two coatings being in series, but due to the lower statistical probability of failed areas in two sheets coming into line at the same time.

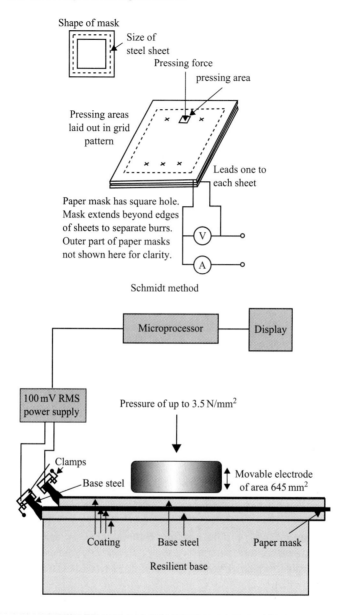

Figure 10.44   Arrangement of the Schmidt tester

The principal use of the Schmidt test is as a type test for insulation to get a view of its probable performance, whereas the Franklin-based tests are more used for quality control purposes.

It is possible to conduct surface insulation tests on steel held at temperatures representative of service conditions in a machine. Clearly this test is much more complex to perform and tends to be reserved for type testing only. Full details of the tests and their associated statistical management of data may be found in the relevant standards [IEC/BS].

10.26   A motor simulation experiment

Because so much controversy came to surround the levels of surface insulation which may be appropriate for steels used in motors, experiments have been undertaken to make a link between insulation figures as assessed by various tests and actual machine performance. It has long been found that it is very difficult to obtain any but the broadest correlation between various surface insulation test methods.

10.26.1   Test programme

A stock of steel was created which had 'very good' insulation. This goodness was arbitrarily decided based on the best material conveniently available. A large pile of sheets was set up such that sheets taken from it could be relied upon to be very close to each other is insulation value. A set of similar 'piles' was made for sheets of 'good', 'fair', 'poor' and 'no insulation'. Again these were arbitrarily chosen, but proved to be in correct rank order under test. A series of insulation test systems was set up covering the Franklin methods A and B [10.12], the British BS 6404 tester, the Schmidt tester, etc.

Sheets from the 'very good insulation' pile were applied to every tester and the results noted. This was repeated with steel from each of the other piles. A large number of sheets from each category were applied to each tester to give statistical coherence to the data. Tests were duplicated before and after annealing treatments as appropriate.

It had been intended that motors would be built from steel drawn from each 'pile' and their performance evaluated to see how steel of varying surface insulation affected this. However, it seemed unlikely that a clear indication could be found when building a realistic number of motors because of the way in which the motor building process affected the behaviour of steel. Figure 10.45 shows (notionally) how power loss varies with each process of manufacture, including stamping, annealing, etc. The degree of 'lift' and 'spread' which occurs makes teasing out the specific effects of coating insulation difficult.

Consequently it was decided to use not real motors but 'simulated motors' in which stacks of steel rings would be used. Such ring stacks could be cut from steel tested for insulation performance and then analysed magnetically. Figure 10.46 outlines the test programme.

Rings were stamped out with an outer diameter of 163 mm and an inner diameter of 90 mm. This corresponds to the back-iron of quite a large lamination. These ring stacks or simulated motor stators were assembled and wound with magnetising and *B* sensing conductors. In this form the power loss of the ring could be examined accurately, see Figure 10.25.

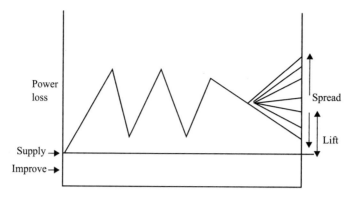

Figure 10.45   'Lift and spread' of magnetic properties of steel through the process of motor manufacture

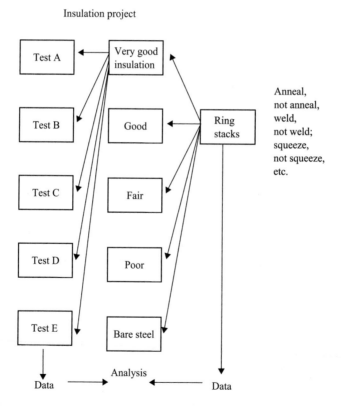

Figure 10.46   Programme of sample analysis

Hydraulic pressurising apparatus for welding the laminae

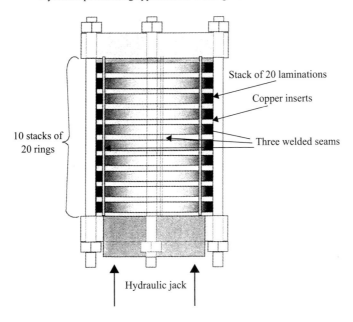

Figure 10.47   Squeezing and welding of rings

Ring stacks were treated in a variety of ways: annealed, not annealed, welded as in a motor stack, hydraulically squeezed to various pressures, etc. In all some 500 ring stacks were evaluated. Figures 10.47 and 10.48 show the form of rings and squeezing method.

10.26.2   Findings

It became clear that the quality or even the presence of a specific insulant had little effect on the iron losses encountered. This was very surprising. In many cases loss variation of the same order of magnitude as the reproducibility of the loss tester are shown in results. This is possible as one data point represents the mean of many measurements.

Figure 10.49 shows results from a range of coatings and it is clear that using 'thick' coatings as a reference condition, bare metal performs as well as thin or medium coatings, the whole range of variation being a very few percent. This result applied for stacks welded using a weld bead down the outside of the stack. Even when paper interleaves were used to provide a 'perfect' coating this proved to behave in a very similar manner to a bare uncoated stack.

If, however, a stack was welded down the inside of the centre hole as well as the outside the losses rose by 300–400%. Figure 10.50 shows that when solid conduction paths were available eddy currents would flow and raise losses.

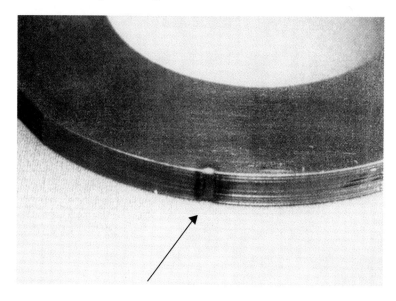

Figure 10.48   Weld on sample stack

In general similar effects of 'bare' versus 'insulated' were seen on unannealed rings suggesting that even the natural oxide on steel as produced was almost as effective in restraining eddy currents as formalised coatings. For this reason the relationship between insulation measurement results and 'motor' performance could not be made in a meaningful way.

Further tests were made using real laminations having stator teeth. In this case results were rather different. Using paper interleaved stacks as a base reference, losses are very similar with the paper omitted. However when welded (down the outside) the losses rise by about 25%. This behaviour differs from that of plain rings, and it was considered possible that burrs on the teeth (the region most liable to burr formation) could, in conjunction with the outside-of-ring weld, give some complete circuits.

Some comparisons were made between laminations stamped on a newly fettled tool and on a worn tool, and there was some evidence that a worn tool giving bigger burrs contributed to higher losses. However the presence of a coating helped reduce losses, suggesting that burrs could 'hide' by digging into the coating below them (Figure 10.51).

10.26.3   Lamination width

Since it appeared that burr management had the greatest effect on small motor laminations some measurements were made in a single sheet loss tester to see at what width and interlaminar emf significant eddy currents could flow across face-to-face steel contact.

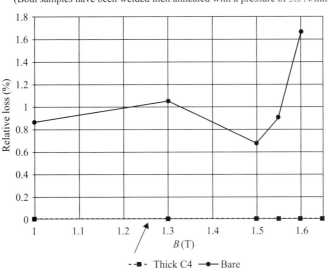

Figure 10.49   Loss results with and without coating

Pairs of samples, one above the other, were tested for both the coated and uncoated state covering the width range 82–500 mm. Above 80 mm some small extra losses could be noted on uncoated sheets. Such widths involve interlaminar emfs typical of quite large machines. Figure 10.52 shows this.

10.26.4   Specially clean surfaces

To check on the possible effect of naturally formed surface oxide layers, some uncoated ring samples were tested before and after exposure to chemical cleaning involving the use of an acid bath. The effect of this extra effort to create a coating-free material was minimal even though loss testing was performed quickly before spontaneous oxide could reform to any great degree.

10.26.5   Annealing atmospheres

Motor stator stacks are annealed after construction, and it is important that such an annealing cycle is evaluated for its effect on stack performance. The annealing cycles given to the samples in the ring tests described involved the careful use of a decarburising (wet hydrogen) atmosphere. For this work the atmosphere was retained down to approximately 200 °C. No noticeable sticking together of laminations was observed.

However, in the interests of speed and economy stator stacks can be exposed to air too early (e.g. 500–600 °C) and then sticking between laminations will occur. If

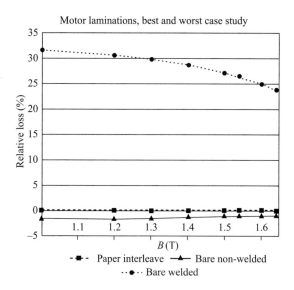

Figure 10.50   Effect of a weld applied to the inside and outside of a stack

an appropriate pre-applied coating is present it will largely restrain the incidence of sticking.

Experiments which allowed premature ingress of air to the annealing process, Figure 10.53, showed that the power losses of test ring stacks then rose sharply, and that the degree of rise was much greater for 'uncoated' rings. This suggests that the formation of obviously coherent adhesions between surfaces is facilitated by the presence of air during the later part of an annealing cycle. More work to examine the exact mechanisms of adhesion formation would be valuable.

10.26.6   Motor repair

It has been reported that motors rewound after repair involving the 'burn-out' of winding fixatives in an autoclave, show higher losses, and that these losses can be reduced by an appropriate choice of the surface coating used in the machine when first built. This suggests that air may have been insufficiently excluded from the burn-out autoclave.

10.26.7   Implications

The studies described suggest that surface coatings on stator steel can help to reduce stray losses, but that the mechanism may involve burr accommodation rather than the restraint of face-to-face interlaminar electron flow. When interlaminar emfs are such as those found on wide lamination back iron (e.g. over 100–200 mm) definite conduction current losses appear. In many cases, stamping and tool life is enhanced

Comparison between 'burred and non-burred' motor lamination

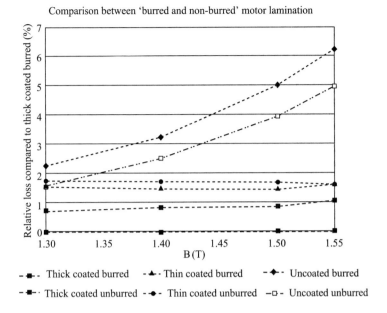

Figure 10.51   Effect of burrs on test samples

by the presence of coatings, which can also prevent sticking of laminations during anneal and facilitate the smooth 'skewing' of rotor stacks.

So, while insulation coatings on motor steel play an important part in facilitating the efficient structuring and operation of the machine the ohmic values obtained from traditional tests may be of limited use in characterising their overall merit. It seems that the presence or absence of an interpenetrative state of the crystal structures of two laminations (at at least two points) is needed to promote high eddy current losses for machines up to medium size [10.13].

10.26.7.1   Thickness

Thickness is a vital parameter of electrical steel. Its evaluation is considered in Chapter 13.

10.26.7.2   Mechanical properties

For many purposes the mechanical properties of electrical steel are as important as magnetic properties. Some relevant properties may be listed as:

- Hardness
- Waviness
- Ductility
- Punchability
- Proof-stress

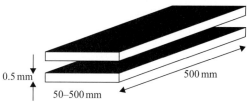

Simulation of larger 'back iron' widths using single sheet tester
Pairs of sheet samples, 50–500 mm wide by 500 mm long, of uncoated and
coated steel were magnctically tested in a single sheet tester

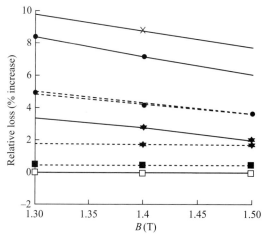

Relative loss difference on various widths of coated and
uncoated steel

■ uncoated, 82 mm	● uncoated, 320 mm
□ coated, 82 mm	● coated, 320 mm
✳ coated, 163 mm	✕ uncoated, 496 mm
✳ uncoated, 163 mm	● coated, 496 mm

Figure 10.52   Effect of sheet width on the impact of insulative coating

- Camber, flatness and edge-drop
- Ultimate tensile stress
- Young's modulus
- Roughness
- Internal stress
- Dynamic friction
- Burr.

10.26.7.3 Hardness

The hardness of a metal is commonly expressed in terms of the degree of permanent
penetration of an indenter into the metal under a specified load. Testers using a ball

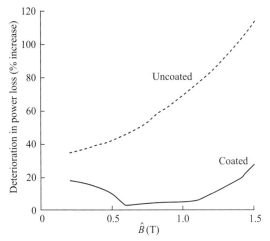

Deterioration in power loss arising from adhesions produced during imperfect anneal

Figure 10.53   Effect of early ingress of air into an annealing chamber

indenter are the Brinell and Rockwell B types. The Vickers hardness tester uses a pyramidal diamond as indenter. This test is widely used for evaluation of electrical steels. The indenter is forced into the metal surface under a fixed load then the diagonals of the indentation are measured via a microscope and tables used to compute the relevant hardness figures taking account of the magnitude of the fixed load.

For thin materials a light load is appropriate and for electrical steel a 2.5–10 kg load is typical. The result of a Vickers hardness test is quoted as a number, e.g. VPN_{10} 140. This stands for Vickers pyramid number 140, and the subscript indicates the fixed load used in kg.

Electrical steels fall into the range 80–200 VPN. 80 VPN represents a fully annealed very soft steel and 200 VPN a strongly temper reduced steel prior to final anneal.

In the USA and Europe lamination stampers prefer VPN values around 140–160 for convenience of stamping. In the Far East softer steel around 100 VPN is happily stamped. Conservatism in the industry has prevented the full exploitation of magnetic possibilities because a particular (and incompatible) range of hardness has been desired.

Unfortunately the Vickers test, while accurate and reproducible, is slow and tedious to perform. A range of electronically controlled rebound testers (based on the vigour with which a hard ball rebounds from a surface) are becoming available which produce a hardness reading directly in digital form for computer storage. These have not as yet found very wide usage in the electrical steels industry.

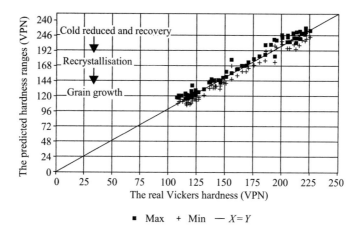

Figure 10.54   Correlation between indenter and magnetic assessment of hardness

Hardness checks are most usually performed on steel samples which have been taken for magnetic testing. In the expectation that magnetic assessment will be universally carried out continuously on-line, efforts have been made to measure hardness continuously on-line.

It has long been known that magnetic properties alter with hardness. Particularly coercive force increases in line with VPN. Attempts in the past to use coercive force as the basis of an on-line hardness test have been frustrated by the loose nature of the correlations found. The need for comprehensive on-line magnetic measurement has led to modern comprehensive on-line testers being available that are able to simultaneously give continuous measurement of coercive force, power loss permeability and remanence.

A wide-ranging study has shown that while no one of these parameters alone has an exact correlation with hardness, a describing algorithm can be created which enables hardness to be accurately determined from a combination of these magnetic features. This is described in [10.14]. Figure 10.54 shows the sort of correlation obtainable and Figure 10.55 shows plots taken along the length of a steel coil in production. Of course this sort of continuous measurement allows variation along a coil to be noted so that the misfortunes of aberrant hardness are readily detectable.

The technique could be applied to steels other than electrical, but of course would require an on-line magnetic measuring device as an extra cost item whereas this is required in the ordinary way for electrical steels.

10.26.7.4   Ductility, proof stress and ultimate tensile stress

These quantities may be described briefly as follows.

Ductility: the percentage of plastic extension experienced by steel under tension up to the point of fracture.

(a)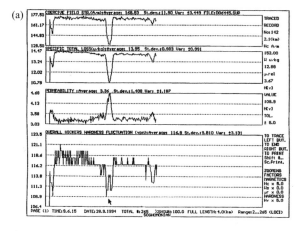

(b)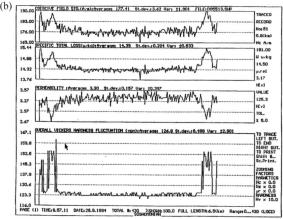

Figure 10.55   Correlation between magnetic properties and hardness

Proof stress, $N/m^2 (= Pa)$: When steel is stretched under a tensile load the extension is elastic up to a 'yield' point beyond which plastic deformation occurs which is not reversible when the load is removed. This point is not easy to determine exactly so it is usual to indicate the load required to produce (conventionally) 0.2% of permanent extension.

Ultimate tensile strength (UTS), N/m^2 *(Pa)*: This corresponds to the stress required to produce fracture in tension. These properties are determined using a tensometer which applies controlled loads and records the amount of stretch arising. Figure 10.56 shows a view of a tensometer and its jaws with a typical electrical steel strip sample in place.

Figure 10.56   Tensile test machine (Instron)

Importance of proof stress, UTS and ductility

Electrical steels are not normally expected to carry high mechanical loads except for those specially strong grades (e.g. Tensiloy) used as load-bearing parts of large rotating machines. However, the business of stamping laminations out of strip requires a close knowledge of these properties. As has been said Europe and the USA is more committed to punching harder steel than is common in the Far East. In either case the properties of steel which make stamping easy and incur low tool wear are reasonably well understood.

Nevertheless up to a point, stamping can be considered to be a 'black art'. The steel properties of relevance are:

(a)   Hardness
(b)   Ductility
(c)   Yield or proof stress
(d)   Yield/UTS ratio
(e)   Presence of a coating and its nature

(f)   Flatness
(g)   Internal stress
(h)   Grain size.

It is beyond the scope of this discussion to examine all the details of punch work, but a few points can be made:

(i)   It is important to match the punch/die clearances to the type and thickness of material being punched. The shearing operation occurring in a die involves some stretching and some fracture – these need to be appropriately balanced.
(ii)   Punch/die lubrication requires careful management.
(iii)   The combination of the punching operation and the presence or absence of residual stress in the steel feedstock has to be managed so that laminations with (for example) round holes to be punched come out with truly round holes rather than any unwelcome degree of ellipticity. Flatness of course is important such that punched round holes remain so in an assembled machine stack. [11.20] discusses punching.

10.26.7.5   Young's modulus

This is the ratio of stress/strain:

$$\frac{f/A}{\Delta s/s}$$

where f = force, A = area, Δs is extension, s = length. Young's modulus, the 'elasticity' of steel, is important if steel components are 'pressed' into position and operate under stress. The stress degrades magnetic properties so the dimensions of deformation need to be known.

Magnetically anisotropic steels such as grain-oriented steel have an anisotropic Young's modulus. This means that stresses in different directions are met with different dimensional deformation rates. In grain-oriented steel Young's modulus in the $0°$ and $90°$ directions to the rolling direction may vary by as much as $2:1$.

As is well known the velocity of sound in a solid is related to Young's modulus. This can have importance in large rotating machines if radial forces take a time to travel from one part of the machine to another in just the time in which flux reversal is required. A knowledge of Young's modulus can guard against unrequired resonances.

Typical guideline Young's modulus figures for electrical steels are:

Grain-oriented in rolling direction: 1.10×10^5 MPa.
Grain-oriented at $90°$ to rolling direction: 1.93×10^5 MPa.
Unoriented steel: 2.0×10^5 MPa.

10.26.7.6   Determination of Young's modulus

Sometimes stress and strain figures from an extensometer may be useful but precise special methods are often preferred. Two major ones are:

(a)   *Static loaded beam*
       Here a 'beam' of electrical steel (an Epstein strip is convenient) is placed on knife edges and progressively loaded at its centre, see Figure 10.57. The deflection

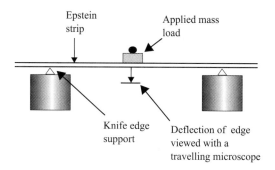

Figure 10.57   Young's modulus measurement by beam loading

of the centre is observed and recorded with a travelling microscope. A plot of deflection/load can be used to get an accurate regression. Then

$$\text{Depression at centre} = \frac{\text{wt applied} \times \ell^3}{\text{Ymod} \times w \times 4t^3}$$

using appropriate units, where

ℓ = spacing between knife edges
w = width of strip
t = thickness of strip.

(b) *Vibrating cantilever*
In this technique a tongue of steel is cut out (6 mm × 200 mm is convenient) and nipped in jaws which are adapted so that two differing depths of penetration into the jaws are possible such that two alternative 'cantilever' lengths are available. When clamped in position the cantilever is excited into flextural resonance by pulse magnetising currents fed to the electromagnet E (Figure 10.58). The frequency of excitation is adjusted until resonance is observed via the microscope. White chalk scraped onto the end of the sample facilitates observation. The operation is repeated for the second value of cantilever length, so that two resonant frequencies f_1 and f_2 are obtained.

By solution between these two, the end effects of clamping are eliminated. The appropriate equations are

$$E = \frac{38.33l^4 \rho n^2}{t^2}$$

where

l = spacing between knife edges
ρ = density in g/cm^3
n = lowest resonant frequency
E = Young's modulus in dynes/cm; (Pa.)

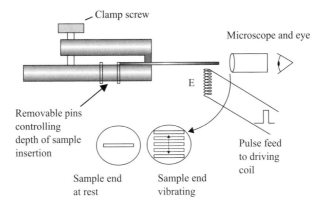

Figure 10.58   Dynamic Young's modulus measurement

and

$$E = \frac{38.33(l_1 + x)^4 \rho n_1^2}{t^2} = \frac{38.33(l_2 + x)^4 \rho n_2^2}{t^2}$$

n_1, l_1 and n_2, l_2 being the two resonant lengths at the two frequencies, and $x = $ end correction.

Hence

$$x = \frac{l_2 n_2^{1/2} - l_1 n_1^{1/2}}{n_1^{1/2} - n_2^{1/2}}$$

This correction can then be appropriately applied.

It is well known that the Young's modulus of steel is affected by its state of magnetisation, and for this reason the pulses of current fed to the electromagnet should be very short compared to the time of a cycle of oscillation, and their amplitude kept as low as is appropriate to secure a clear resonance.

10.26.7.7   Dynamic friction

As has been discussed, electrical steels often carry an insulating coating. Users have varying opinions as to the frictional properties they would prefer for these surfaces. Aesthetically a bright shiny smooth surface is attractive in appearance, but is not always appropriate.

(a)   *Low frictional surface*
When stacks of laminations are skewed for the final shape of a stator stack it is beneficial if the laminations slip easily over each other. Oils used in punching could facilitate this. Also transformer steel involves the precision cutting of core plate lengths at high speed. Some machines favour low friction, allowing rapid sliding of strip on strip. Other machines rely on inter-strip friction to slow them down.

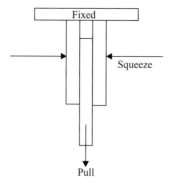

*Figure 10.59   Friction test. The force required to pull strip out under a set squeeze
load indicates frictional properties*

(b) *High friction*
When a stack of laminations is built up to a considerable height and clamped,
the stability of such a stack can depend on the surface-to-surface frictional
properties of the steel. In some cases minimum coefficients of friction may
be required, Figure 10.59. Depending on machine design, high or low friction
may be favoured.

10.26.7.8   Roughness

The roughness of the surface of electrical steels is an important parameter. Motivations
for high or low roughness may be cited as follows:

Low roughness is needed for low coefficient of friction – see above. It improves
magnetic performance by avoiding unfavourable closure domain structures at
the surface, and magnetic reluctance between one prominence and another. The
stacking factor of the steel is impaired if rough.

High roughness. This is useful for allowing gas penetration between layers of
metal being annealed in a stack, especially when decarburisation is needed. Here
gas has to get to all surfaces and reaction products removed. The creation of steam
blueing as a surface insulation coating also needs good access to metal surfaces.

It has been found that an increased surface roughness reduces the tendency for
laminations annealed in a stack to stick to each other. Of course careful management of atmosphere during cooling (exclude oxygen until below 300 °C) helps
avoid sticking, but roughness in the range 1–2 micron (40–80 µin) also makes
sticking less likely.

Scales of roughness. Methods of test: Figure 10.60 shows a commercial roughness
tester.

Figure 10.60 Commercial roughness testing machine

10.26.7.9 Internal stress

Residual stress within steel is always detrimental to magnetic properties and can have various forms:

Stress arising from rolling

Normally steel is annealed after rolling before being used, but it may be stamped in the partly hard state after temper rolling and before anneal. The stress associated with the plastic deformation which occurred during rolling may show itself in various ways – stamped laminations may emerge with slightly oval holes following stamping with a true round punch and die. Flat steel as slit may develop waviness as locked-in stress redistributes itself.

Bending/coiling stress

If strip is coiled over too tight a diameter it can develop so-called 'coil set', that is a slight permanent curvature similar to the coiling shape, showing that the elastic limit had been exceeded during coiling or passing over rollers. If the proof stress of the material is known, and the radius of bend applied, it is soon clear if the outer filaments of metal have experienced plastic deformation.

Stretch

Too high a tension in a process line may pull a strip so hard that slight permanent stretch is produced. If the 'shape' of the steel was imperfect before stretch it may improve as 'waves' are pulled flat. In doing so permanent stress may become inbuilt.

The temperature at which stress is experienced is also important as the yield point reduces with rising temperature.

It has been pointed out that the stamping process can also introduce stresses able to produce distortion in a finished component. Further, residual stress in steel may in fact help to reduce final distortion or may exacerbate it. The steel-producing industry finds it best to aim always to produce stress-free material.

In the case of fully finished steel, stress is sometimes seen as being expressed by how much the magnetic properties of the steel can be improved by a thorough annealing treatment. If the improvement of a magnetic parameter, e.g. power loss, is noted to be 2% it is convenient to say 'the stress level in it was 2%'. This is not a very scientific measure but has its usefulness.

The exposure of so-called fully finished steel to bending and stretching forces need not always be detrimental. If an anneal is applied to relieve punching stresses so that magnetic properties may be optimised, then any locked-in pre-stamping stress may actually aid grain growth during the post-stamping anneal and provide further enhancement of properties. Unless material is constrained flat during anneal, stress relief may lead to extra distortion.

Measurement of stress

The magnetic measurement of 'stress' is comparatively straightforward. Properties may be monitored before and after an annealing treatment. A piece of flat steel may or may not contain a system of locked-in stress and various efforts have been made to make such stress apparent.

Comb test

In this test a sample of steel is first painted with an etch-resistant varnish. Then lines are cut through the varnish to expose bare metal as shown in Figure 10.61. Photo-resist methods can also be used. Finally the metal is immersed in an acid bath (e.g. at 100 °C for 15 minutes). This etches right through the metal, producing 'comb teeth' in two directions. The teeth edges are not now attached along their length and remaining stresses can become unbalanced and deform the teeth. So far purely empirical assessments of tooth deformation have been used to give findings of stress levels as 'good', 'poor' or 'bad'.

The knowledge that such tests are possible has on occasion been found to improve the concentration of setting-up operators on a slitting line where optimum set-up minimises the introduction of stress.

10.26.7.10   Burr

Inevitably the punching or shearing of steel will produce a burr. Figure 10.62 shows an example of burr. The skill of the punching and shearing designer controls how

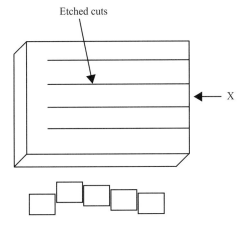

Etched cuts

X

View looking at ends (X). Comb teeth deflect as internal stress is released

Figure 10.61   Etching of cuts to form a 'comb' reveals locked-in stress

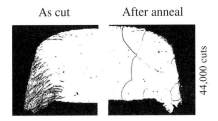

As cut   After anneal

44,000 cuts

Figure 10.62   The form of burr which develops as tools wear (see also Chapter 15)

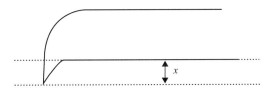

x

Figure 10.63   Burr height defined as x. x increases as tools wear

good a result is obtainable from a given type of steel. In general if steel is soft it tends to smear out into a long burr; if hard, it cuts/fractures more cleanly. The whole area of slitting and stamping is complex and has at present elements of both art and science.

It is sufficient here to consider how burr can be assessed. Clearly burr that contributes to short circuits between laminations is unwelcome. Figure 10.63 shows the general form of a burr. x is considered to be the burr height and is expressed in microns.

The burr is delicate and if measured with a micrometer there is a possibility of flattening the burr with the micrometer jaw. A special flatness table is often used with a micrometer suitably mounted. Attempts have been made to use a slowly advanced micrometer jaw set-up so that as soon as electrical contact is made with a burr peak a reading is taken. The delicate top of the burr may or may not be considered significant to users. Multiple measurements along an edge are made to give a statistical expression of burr.

Burr flattening

Some slitting operations use special, post-slit, burr flattening/grinding rolls that effectively remove the major part of burr. The application of gas heating jets to the edges of strip can be used to 'burn' the very thin burr edge and reduce its potential to cut into coatings [10.15].

10.26.7.11   Flatness

It is desirable that electrical steel shall be flat. The major determinator of flatness is the cold rolling mill. It is not within the scope of this discussion to examine the whole art and science of cold rolling. Sufficient to note that flatness is related to:

Roll shape
Back tension
Coiling tension
Lubrication – general and across the strip width – differential
Degree of reduction
Strip temperature
Rolling speed.

All of these may be controlled and can be part of a real-time computer control system aimed at giving output strip which is both flat and of exactly the desired thickness. Where the rolling concerned is aimed at providing the final strip condition, best flatness is desirable.

When the rolling is providing steel as input to other continuous process lines deviation from flatness may help the control of strip in long lines.

A range of commercial systems for flatness assessment is available. During rolling the term used is not flatness but 'shape'. This term is descriptive of the length of notional metal filaments at different positions across strip width. Typical conditions could be described as 'wavy edge', tight edge, loose middle, etc. Asymmetric patterns of shape are always unwelcome.

A unit of shape is the Mon which is characterised by a 'filament' length of 0.01% greater or smaller than the overall mean (derived from monstrare, to show or make evident).

Under the tension which is usually applied during rolling, the steel normally has a flat appearance and 'shape' is latent until the tension is relaxed.

In commercial terms waviness is characterised by a wavelength and a wave height. Product standards specify limits to these quantities. A more recently used measure of shape is the 'I unit' corresponding to 0.001% length differential.

Strip is annealed in a continuous annealing furnace maybe hundreds of metres long, perhaps at 800 °C. It must be cooled to a temperature of the order of 100 °C before emerging. A useful device for cooling is a gas jet cooler. This device blasts the hot strip with cool gas so that its temperature falls as required.

Careful control of such cooling is needed so that the 'shape' and internal stress of the steel is not adversely affected by edges cooling faster than the centre, etc.

If the atmosphere were decarburising and of a high hydrogen content, then the mass of gas to be moved is comparatively low compared with its ability to take up heat. Moles of gas take up equal amounts of heat for equal temperature rises. One mole of hydrogen (22.4 litres) weighs only some 2 grams whereas a mole of nitrogen weighs about 28 grams.

Camber

The stress condition of steel strip can lead to its taking up, especially during slitting, a cambered shape. Figure 10.64 shows this. Strict limits are imposed on the degree of camber which is considered tolerable and careful management of slitting can minimise its incidence.

Edge drop

There is a built-in tendency for rolling mills to impose a profile on strip other than an exact rectangular cross-section. The principal unwanted features are 'crown' and edge drop. Variation of thickness with width is detrimental to the production of evenly shaped stacks of laminations. Figure 10.65 shows crown and edge drop, and illustrates the effect on a lamination pile cut from an edge slit. Of course the edges could be discarded, but the cost of metal lost as scrap would then escalate. Strenuous efforts are made by the use of specially shaped rolls, slightly crossed rolls and other means to create steel free from edge drop.

The progressive rotation of laminations as a stack is built would not only even out the effects of edge drop, but equalise the remaining magnetic anisotropy within

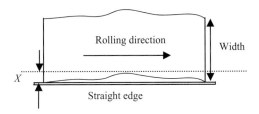

Figure 10.64   Camber of rolled sheet (exagerated scale). The distance X which is apparent when a 1–2 metres long sheet is offered to a straight edge is indicative of camber

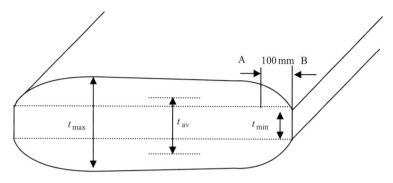

Crown = $t_{max} - t_{av}$. 'Edge drop' is the thickness change over the edge region A–B. The thickness variations are shown on a greatly exaggerated scale.

A lamination pile cut from steel with taper will not pile level unless individual laminations are rotated

Figure 10.65   Edge drop and crown

nominally non-oriented steels. Appropriate domestic and international standards set acceptable limits on edge drop and crown.

Crown and edge drop can be minimised by selection of the profiles of hot rolling rolls or by operating pairs of hot rolls in a slightly 'crossed' mode in which the effective roll bite is more open towards the edges of the strip.

10.26.7.12   Dirt

In former times various parts of the production process for electrical steels could lead to a thin film of dirt clinging to the strip surface. This could contain fragments of the decomposition products of rolling oil, micro-flakes of annealing scale, and so on.

Progressively dirt has been eliminated and it can be monitored by the use of adhesive tape and reflectometry. If transparent cellulose adhesive tape is pressed onto a strip surface then removed, any micro-particles of 'dirt' come away with it. Placing the same tape onto white card, where it adheres, allows dirt to be examined (Figure 10.66).

For numerical objectivity, the optical reflectivity of the card and tape without a prior application to a steel surface may be compared with the after-dirt-lift-off reflectivity.

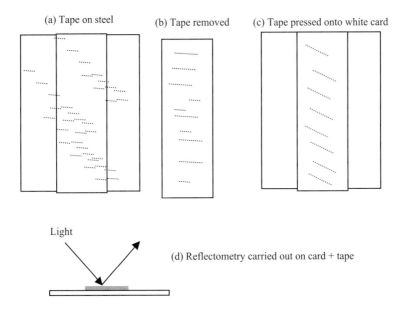

(a) Tape on steel   (b) Tape removed   (c) Tape pressed onto white card

Light

(d) Reflectometry carried out on card + tape

Figure 10.66   Dirt assessment

10.26.7.13   Punchability

This is normally expressed as the number of punching operations performed by a standard punch and die on a given variety of steel before a measured degree of burr appears on the cut edges.

Punchability varies with:

- Steel composition, e.g. percentage of silicon and phosphorus
- Coating: phosphate coatings are aggressive; organic ones help stamping
- Die lubricants
- Punch/die design
- Hardness of steel.

These all interrelate.

Magnetic performance must always be considered in the light of the effect on punchability of the means used to achieve it. Equivalently, stamping houses are challenged to stamp new steels economically, when these have been evolved to optimise magnetic properties.

10.26.7.14   Thermal conductivity and resistivity

The bulk resistivity of steel follows closely the percentage of silicon present. Figure 10.67 shows this relationship. This relationship is sufficiently close for resistivity to be used as a substitute, where convenient, for chemical analysis. Added aluminium behaves very like silicon, see Figure 10.67.

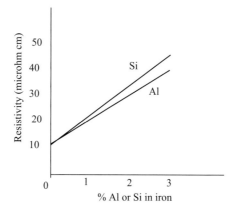

Figure 10.67   Comparison of the effect of Si and Al on the resistivity of iron

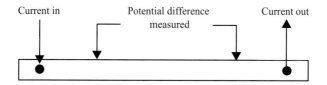

Figure 10.68   A four-terminal resistor

An Epstein strip is a convenient sample with which to measure resistivity. This can be done by using the strip as a 'four-terminal resistor' in which current is fed into the ends of the strip and the potential difference measured at points far enough from the ends for uniform current flow to have been established. Figure 10.68 outlines this method.

A simple method of density and resistivity assessment has been devised by Van der Pauw [10.16] and is discussed in detail by Sievert [10.17].

In Van der Pauw's method a piece of sheet of arbitrary shape can be used for a rapid determination. The method is included in BS6404 part 13, 1996. The work by Schmidt and Huneus also merits study [10.18].

Alternatively a special Maxwell bridge may be used. The resistivity of steel varies with temperature such that alloy-free steel (12 $\mu\Omega$ cm at 20 °C) increases in resistivity by some 0.35%/°C. As is the way with alloys, electron flow interactions mean that, say, 3% silicon steel (50 $\mu\Omega$ cm at 20 °C) rises only 0.1%/°C in resistivity.

These results are useful since they give convenient access to thermal conductivity. In an electrical machine the core laminations are often a significant channel for rejection of heat, due to iron and copper losses, to any coolant employed. Thermal conductivity is then important.

Electrical and thermal conductivities are connected via the Lorenz function such that

$$\frac{\text{thermal conductivity}}{\text{electrical conductivity}} \simeq \text{approx. constant}$$

The ratio is about 3 for the usual system of units. The Lorenz ratio is

$$\frac{\text{thermal conductivity} \times \text{resistivity}}{\text{temperature in kelvins}}$$

This is approximately 3×10^{-8}.

Note, resistivity is the inverse of conductivity, so the ratio shown is in fact the ratio of thermal and electrical conductivities divided by the absolute temperature. The dimensions of the Lorenz ratio are volts squared/kelvins squared.

By use of this relationship the thermal conductivity of steel can be usefully predicted. Of course, a stack of laminations has a much higher thermal conductivity along the length of laminations as compared to that through a stack, Figure 10.69. The through-the-stack conductivity is often some 30–40 times lower than through the metal alone. The effects of surface coatings, roughness and clamping pressure all modify the latter conductivity.

When through-the-stack measurements have to be made, a complex set-up of constant flow thermal conductivity apparatus has to be used involving thermal guard rings and long periods for the system to reach equilibrium. These measurements are not often undertaken.

Typical thermal conductivities are (parallel to plane of sheet):

- Grain-oriented steel   26–27 W/mK
- 1.3% Si steel   45 W/mK
- Silicon-free steel   66 W/mK

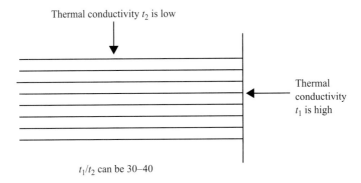

Figure 10.69   Contrast of directional thermal conductivities in a stack of laminations

Conductivity in the normal-to-plane direction in a stack depends on the detail of coatings, clamping pressure and lamination thickness. Values some 30–40 times lower than in-plane values are typical.

The units of thermal conductivity are watts/metre kelvin.

Solvent resistance

Although not exactly a physical test, it is necessary to know the solvent resistance of coatings on steel. Many motors are operated in hermetically sealed systems in which the laminations used are exposed, long-term, to refrigerant fluids. The response of coated steel is usually assessed in respect of test fluids. These are discussed in the chapter on coatings. The evolution of new 'green' refrigerants means that this property of coated steel has to be continuously reviewed.

Rapid per cent silicon test

It has been found that the thermal emf generated at the contact between silicon steel and a neutral metal (e.g. mild steel) is a function of the percentage of silicon in the steel. This fact has been used to develop a probe able to read out the percentage of silicon directly when pressed onto a steel surface. The probe incorporates a heating element to bring the point of contact to around 100 °C. A thermocouple embedded in the probe notes the actual temperature.

A microprocessor takes in data about temperature and developed thermal emf and produces a reading in percentage of silicon. A 'cold' junction consists of a clip to some other part of the steel being examined and completes the circuit. Ref. [10.19] describes the system.

10.26.7.15　Thermal expansivity

Sometimes the tightness of fit of motor stators into casings can be influenced by the thermal expansivities of the material of the case compared with the stator steel. Thermal expansivities of electrical steels are of the order of 12×10^{-6} per K.

10.26.7.16　Stacking factor (space factor)

A stack of Epstein strips may be squeezed at $1.0 \, \text{N/mm}^2$ pressure and the stack thickness noted from a dial gauge. From this figure and the sample mass (and number of strips) the percentage effective space occupancy by metal can be calculated. The result is expressed as a percentage figure.

References

10.1. GUMLICH, E. and ROSE, P.: *Electrotechnische Z.*, 1905, **403**.

10.2. BURGWIN, S.L.: 'A method of magnetic testing for sheet metal', *Rev. Sci. Inst.*, 1976, **7**, p. 272.

10.3. *ASTM Symposium on Magnetic Testing* (1948) Special publication No. 85, ASTM, p. 167(c).

10.4. NORMA U-function meter, instruction manual, Norma Messtechnik GmBH, Vienna.

10.5. DANNATT, C.: 'Energy loss testing of magnetic materials utilizing a single strip specimen', *J. Sci Inst.*, 1933, **8**, pp. 276–85.

10.6. BECKLEY, P.: 'Continuous power loss measurement with and against the rolling direction of electrical steel using non-enwrapping magnetisers', *Proc. IEE*, 1983, **130** (6), pp. 313–21.

10.7. BECKLEY, P. and LODGE, J.: 'A magnetisation system for a thin steel flaw detector'. *British Journal of Non Destructive Testing*, 1977, pp. 19–20.

10.8. GOLDING, E. W.: 'Electrical measurements and measuring instruments' (Pitman, London, 1949).

10.9. BOON, C. R. and THOMPSON, J. E.: *Proc. IEE*, 1965, **112**, p. 2147.

10.10. EWING, J. A.: *IEE Proc. Inst. Electr. Eng.*, 1895, **24**, p. 398. Also STARLING, S. G. and WOODALL, A. J. 'Electricity and magnetism' (Longmans, 1953), pp. 284–5.

10.11. SCROGGIE, M. G.: 'The design of iron core chokes', *Wireless World*, 1 June 1932, pp. 558–61.

10.12. BS 6404 Pt 11 1991.

10.13. BECKLEY, P., *et al.*: 'Impact of surface coating insulation on small motor performance', *IEE Proc. Elec. Pow. App.*, 1998, **145** (5), pp. 409–13.

10.14. SOGHOMONIUM, Z. S., BECKLEY, P. and MOSES, A. J.: 'On-line hardness assessment of CRML steels', *Amr. Soc. Materials Conf.*, Chicago, Oct 1996.

10.15. CARLBERG, P. M.: 'The cutting of electrical steel sheet', Jernkontoret Punching Conference, Stockholm, 9 November 1971.

10.16. VAN DER PAUW, L. J.: 'A method of measuring specific resistivity and Hall effect of discs of arbitrary shape', *Philips Res. Repts*, 1958, **13**, pp. 1–9.

10.17. SIEVERT, J.: 'The determination of the density of magnetic sheet steel using strip and sheet samples', *J. Magn. Magn. Mater.*, 1994, **133**, pp. 390–2.

10.18. SCHMIDT, K. H. and HUNEUS, H.: 'Determination of the density of electrical steel made from iron–silicon alloys with small aluminium content', *Techn. Messen.*, 1981, **48**, pp. 375–9.

10.19. Richard Foundries Ltd., Phoenix Iron Works, Leicester, LE4 6FY.

Additional reading

HAGUE, B.: 'Alternating current bridge methods' (Pitman, London, 1938).

Chapter 11

Cost and quality issues

Steel metallurgists find great satisfaction in producing top-quality steels with first-class magnetic properties. Machine designers, similarly, wish to produce top-quality machines. Unfortunately both are bound within economic restraints to produce products which are marketable and profitable. For the steel-maker compromises have to be made in the direction of reduced costs. This usually involves:

- Choice of steel composition
- Choice of rolling and heat treatment regimes
- Choice of coatings.

The machine builder wishes to use inexpensive feedstock and fabricating techniques yet produce motors/generators which are:

- Energy efficient, with low iron and copper losses
- Small in weight and size, having high specific horse power
- Able to operate over a wide speed range
- Offering desirable starting and operating torque characteristics.

Suppose a machine were aimed at a low-loss performance. A low-loss steel may be selected; this may be *thin*, e.g. 0.35 mm, and of *high alloy* content, e.g. 3.2% silicon. The thin steel involves more stampings and impaired space factor.

If high-silicon steel is used machine behaviour will be best in the medium range of working inductions up to 1.5 T. Working at a moderate induction involves a higher cross-section of steel to produce a given flux output. A higher cross-section requires a longer copper route to encircle it, raising copper losses. Thus the machine will be large and heavy. However, both the thin steel and its raised resistivity mean that high-frequency operation will be good.

Increased electrical resistivity goes hand in hand with reduced thermal conductivity, so the removal of heat for cool running will be more difficult.

Steel coatings may or may not be used depending on whether machine size (interlaminar emfs) or punching practice (burr size) make a formal insulation desirable.

In general to have low loss a machine needs thin resistive steel and a design which best exploits it.

Suppose, however, that the lowest cost were the objective. Then a thick steel 1.0 mm thick, unalloyed, and uncoated could be used. Copper losses will be reduced by cores of small cross-section (short copper route round a small cross-section of steel), if operation at $\hat{B} = 1.8 + T$ is required, but high magnetising currents will themselves contribute to power loss via copper loss (low permeability at highest inductions). Lowest cost heat treatments and only first-order decarburisation may be used to cut back on steel process costs.

Such a machine will be cheap, small and light and will run hot. Clearly it is suited to low duty cycle operation.

If a machine must be as small and light as possible (e.g. aerospace) then operation at high flux densities and high frequencies is needed. For this, very thin steel, e.g. 0.1 mm, is needed either of 3.5% silicon content or use may be made of specialist cobalt alloyed steel. Where high temperature survival is vital (jet engine peripherals) inorganic surface insulation coatings and cobalt alloys are appropriate.

Within the electrical steel-producing route economies can be secured by:

- Specifying a steel composition in common volume production (no 'specials' cost)
- Omitting decarburising and desulphurising at steel-making
- Omitting a homogenising anneal at the 'hot band' stage
- Deciding how much strand decarburising effort to apply
- Choosing box or strand annealing. Either one may be the most economical in a given case
- Using a critical grain growth process or not
- Applying a coating, or not. If one is used what type should it be?

Obviously, the fewer distinct types of steel a producer makes the lower will be the unit cost of each.

To support production of a 'special' or a new variety requires careful market evaluation and co-operation with one or more potential users. Very often a machine producer will procure laminations from an independent stamping house and there will be no direct contact between steel producer and motor builder. In such a case the stamping house will wish to base its operation on grades of steel commonly available round the world and which are closely specified. This level of flexibility can exclude the machine builder from the fruits of direct design co-operation between steel producer and motor maker.

It is to be hoped that material design and machine design can be drawn together so that optimised benefit can be drawn from the innovative skills of all parties.

11.1   Actual costs

Many factors influence the market price of steel products. Some of these are set out below.

Any actual figures are of course doomed to date rapidly, but as of 2002 it could be said that semi-finished steel would be found in a bracket around £250–400 per tonne; fully finished steel around £400–800 per tonne, and grain-oriented steel in the range £1000–2000 per tonne. It may be asked why these ranges are so vague. Some of the related factors are:

(a) Chemical composition. ULC? ULS? Each can be applied with more or less completeness adding tens of £/tonne.
(b) Alloyed %Si. Ferrosilicon is expensive and rolling silicon steel is more damaging to rolls.
(c) Thickness. How much rolling is required?
(d) Metallurgy. How much grain texture control is required? Is minimal anisotropy sought, or a particular grain size?
(e) Magnetic grade. Power loss and permeability. The number of rolling and heat treatment stages needed will vary. More extensive treatment gives better results.
(f) Coatings. Simple, complex, heat-resisting or none?
(g) State of steel. Semi-finished or fully finished?
(h) Format. Full coil (e.g. 20 tonne), slit coil sheet, etc. Slitting and punching are expensive operations.
(i) Commercial features. Quantity, delivery rate, payment terms.
(j) Transport to where? FOB at docks or delivered to user's plant?

A parallel could be drawn with motor cars. Small and fuel economical, bigger and sluggish, has everything but costs a lot.

Supply and demand greatly influence costs. A merchant carrying stocks in a warehouse must cover the cost of holding a varied stock, but an agreement between a user needing thousands of tonnes and a steel-maker could involve the creation of special grades and compositions with the benefits of bulk purchase and direct delivery.

Some countries insist that packaging must all be 'green' or be recovered to its country of origin – this is expensive.

A mesh of tariffs and trade controls influence what can be sold where in the world. Electrical Steels are the main carriers of flux serving industry in the world. Special materials such as cobalt–iron alloys can find a place in aerospace applications where weight, bulk and high-temperature operation outweigh price. These materials can command thousands of £/tonne and fall outside the scope of this book.

11.2 Specific applications

The instances described below do not relate to specific currently sold devices but serve to illustrate the points involved.

11.2.1 Fan heater motor

The requirement is for a small motor to drive a fan incorporated in a domestic fan heater. There is no special need for high efficiency as the device is intended to produce heated air by virtue of its function. The motor must be quiet and reliable and develop a modest torque. A shaded pole motor is ideal for this purpose. A fan load will cause the motor to drop below synchronous speed and deliver the torque associated with air drag at that rpm. The machine must be long-lived, reliable and cheap.

The stator will be wound for 230 volts operation, 50 Hz (in the UK) and iron loss must be low enough so that an inexpensive insulated winding does not fail due to overheating, neither do bearings dry out prematurely. The motor will probably be designed not to fail dangerously if its fan is stalled or the device intake/output becomes obstructed. Overtemperature cut outs can be used. So stator steel need not be low loss but should deliver a fair flux density for a modest magnetising current. Probably a 0.65 mm thick semi-finished steel lacking silicon additions or special carbon or sulphur reductions would be used.

- Possible material: 1000 65 D5/M1000 65 D.
- Maximum power loss: 10 watts/kg at $B_{max} = 1.5$ T, 50 Hz.
- 0.65 mm thick, uncoated.

  This material stamps well and benefits from a simple final anneal.

Overview: The core material is cheap, does not need to be of low power loss, and does not need coating. The oxide arising from the post-stamping anneal and the small interlaminar voltage means that special insulation is not needed.

11.2.2   Lathe motor

This could well be a 5 kW, three-phase induction motor with efficiency running up to 80+ %. If fitted as integral to the machine tool it will be purchased as a budget item by the machine tool builder. The lathe user is unlikely to question closely the exact efficiency of the motor. Its duty cycle is intermittent, its load is very variable from off-load up to full-load and the system inefficiency associated with gears, pullies, etc. means that the underlying motor efficiency is not clearly evident. Such a motor could benefit from use of an 890-50-D5, M890-50D, 0.5 mm steel, or, better, one of the Polycor grades. Efficiencies ranging up to 88–90% will come to be expected.

European legislation is beginning to squeeze more tightly on permitted motor efficiency ranges so that better steel is becoming relevant.

Further, if the machine uses power electronics for speed control the motor rating may have to be reduced to accommodate the effects of harmonics in the motor drive waveform, or even pulse operation under PWM.

Suitable steel could then be 570.65D5/M570.65D Polycor type.

11.2.3   Large fan 150 kW

This may be used to drive the fume extraction system of a piece of industrial plant or chemical process. The specifying engineer will have given close consideration to the cost of iron losses in the motor over its useful life and factored this into the longer-term cost of ownership. Efficiency in the 90+ % range will be expected and a fully processed steel containing 1–3% silicon at 0.5 mm may be used. This could be M530–50A grade.

11.3 Spread of quality within a grade

It is always desirable to produce a product with as small a spread of properties as can be managed. However, tightening the range of properties beyond a certain degree becomes prohibitively expensive and involves scrap and wastage.

Consider Figure 11.1 in which the power loss of a steel variety is plotted against frequency of incidence. Clearly everyone wants steel from region L-E, yet if the grade maximum is R and the average is at M, some appropriate marketing scheme is required.

Possible strategies are:

(1) Offer the whole histogram of performance at a particular price, if it all comes, unrobbed of area L-E.
(2) Offer the whole histogram with R chopped off.
(3) Offer L-E at a premium and R at a discount to appropriate customers to whose needs this is fitted.

Clearly, a motor designer will have to consider what spread of characteristics of input steel are tolerable. Designs centred on the modal property, the grade maximum, etc. will require careful thought. Further, the designer needs to be aware of the degradation and spread of properties induced in feedstock steel by the variability built into his production process. Figure 10.45 illustrates the process of lift and spread which occurs during machine production. If lift and spread are great then variation of input material may be of limited importance and vice versa.

A careful study of the ways in which degradation of steel properties can arise and a close audit of their avoidance during motor manufacture can enable a designer to more fully employ the best properties that steel has to offer.

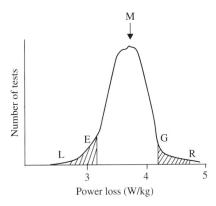

Figure 11.1 Histogram of power loss figures

Close attention needs to be paid to:

- Avoidance of stress in assembly
- Provision of appropriate annealing – especially for ultra-low carbon steel
- Avoidance of punching practices that produce burrs able to give short-circuits between laminations
- Constructive application of rotor face and/or bore skimming so that smeared-over metal does not produce short-circuits.

When motor producers purchase laminations from a stamping house there is less opportunity to match properties to application since the steel sources will be unknown and the standards of quality obscure.

When a motor maker has his own stamping and (if used) annealing plant and purchases steel direct from steel-makers the thorough optimisation of quality and economy is better facilitated.

Chapter 12

Competing materials

Electrical steels perform very well in the rôle of high-quality flux magnifiers available at moderate cost. Some other materials perform this function better in some respects and worse in others. Most of them are much more expensive than electrical steels but have specialist properties which make their use appropriate for some niche applications. Figure 12.1 lists some of these. The spheres of use of these materials and a view of their properties will be briefly examined.

12.1 Nickel irons

Nickel is a very expensive metal, but when alloyed with iron in percentages from 30–80% it produces a range of alloys whose principal differences from iron and iron–silicon are:

(1) Very high relative permeabilities – maximum values up to 10^6
(2) Reduced saturation inductions in the range 1–1.5 tesla

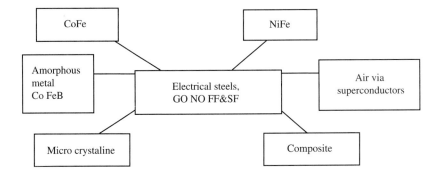

Figure 12.1   Niche applications of soft magnetic materials

(3)   Curie points in the range 300–550 °C
(4)   Coercivities in the range 1–10 A/m.

The very high permeabilities and low coercivities available from iron–nickel alloys make them a first choice for shielding duties. The low power losses and high electrical resistivity offered is useful as is the squareness of the BH loops characteristic of nickel iron.

The reduced saturation induction, lowered Curie point and high cost are the principal deterrents to wider use. Nickel–iron alloys tend to have a high stress sensitivity and must be carefully handled. Where high-performance shielding is needed in the presence of strong fields a two-stage screen is useful in which electrical steel is used as a first-stage flux shunt and a nickel–iron alloy as a 'mop-up' stage. Nickel–irons are much in demand for pulse transformers and like devices, but seldom figure in rotating machine designs.

12.2   Cobalt irons

When cobalt is added to iron the saturation induction of the alloy is raised. Some 2.45 T can be obtained at 25% cobalt. The Curie temperature is then well over 900 °C. Cobalt is an expensive metal so although the advantages of high saturation induction and performance at elevated temperatures are clear, cobalt alloys are only specified where these properties are vitally necessary. The high value of cobalt makes recovery of CoFe scrap from discarded devices worthwhile. Aircraft rotating machines where size, weight and high temperature operation are vital make good use of cobalt alloys.

12.3   Amorphous metals

In general, reduction of power loss and increase of permeability in electrical steels is sought by increasing the perfection of crystal orientation and cleanliness of the metal.

An alternative approach exists. If a suitable alloy is cooled very rapidly, e.g. some million or more degrees C per second, it is possible to achieve solidification without crystallisation. Atoms become locked in place before the energy minimisation associated with regular crystal formation can occur. This structure may be compared to that of ordinary glass in that it is an unstable structure able to form crystals under appropriate conditions.

Figure 12.2 outlines the basic idea of melt spinning in which an appropriate liquid metal is very rapidly cooled to form a thin solid ribbon.

The superior properties of amorphous metal arise from the thinness and high electrical resistivity of the alloy, e.g. 30 microns thick and 100+ microhm cm resistivity which restrains eddy currents, and the fact that domain walls can move freely through the random atomic structure. Makers of silicon steel concentrate on making it as defect-free as possible so as not to pin domain walls. Amorphous metal can be thought of as one huge dislocation mat in which a domain wall has the same energy in one place as in another so that movement becomes free.

(a)

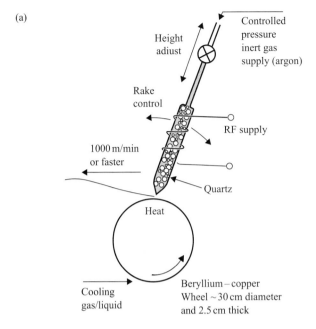

Controlled
pressure
inert gas
supply (argon)

Height
adiust

Rake
control

RF supply

1000 m/min
or faster

Quartz

Heat

Cooling
gas/liquid

Beryllium – copper
Wheel ~ 30 cm diameter
and 2.5 cm thick

(b)

Figure 12.2 *a* Outline of melt-spinning technique for production of amorphous
metal. Industrial production develops this into 30 + cm wide foil
b Amorphous ribbon

The spin casting technique has been developed so that strip more than 30 cm wide
is now made commercially. The rapid cooling results in a strip which is highly stressed
and requires a stress-relief anneal. A low temperature, e.g. 250 °C, must be used for
this to avoid provoking destructive crystallisation of the metal.

The need for a high degree of alloying to permit the amorphous state to be attained
means that the saturation induction of the metal is comparatively low. Usually application of amorphous metal of the iron–boron type is limited to inductions below
$\hat{B} = 1.6$ T.

Although power losses are only some $\frac{1}{3}$ of those for grain-oriented silicon steel, the cost of producing amorphous metal and the problems of handling it and avoiding compressive stress (to which it reacts with increased loss) cause its widespread use for large machines to be limited.

Amorphous metals involving cobalt have properties especially suitable for small inverter cores, but its high cost limits the size of device to which it is usually applied. Attempts have been made to produce a composite material made up of several layers of amorphous metal cohered into a sheet some 0.2 mm thick. This type of core plate which could enable stacked cores to be more readily constructed from amorphous metal is not, however, currently commercially available [12.1].

The stress sensitivity of amorphous materials can be turned to good account in the development of stress sensors.

12.4   Microcrystalline alloys

The melt spinning techniques used for making amorphous material can also be applied to alloys which do not contain sufficient glass formers to make amorphous ribbon. It is possible to use alloys which are not normally rollable and obtain a ductile strip made up of very tiny crystals (microcrystalline) which has favourable magnetic properties. The lack of large percentages of glass formers means that saturation induction is not overly impaired.

Microcrystalline alloys are produced via ultra-rapid quenching just as is amorphous metal. However an alternative process sprays steel in droplet form to make a solid. Here the crystals which form bear a relationship to the droplet size and offer the opportunity to further control crystal size. Spray-formed steel has the potential for rapid production of thin strip provided problems of surface roughness can be overcome.

12.5   Composite materials

The methods used to restrain eddy currents in electrical sheet steel – thinness and raised resistivity – require that solid cores be made up of laminations. It is possible to create a solid core by the aggregation and pressing of iron (or iron alloy) particles such that a thin coating on the surface of particles insulate them from each other and leads to a composite in which eddy currents are largely restrained.

This approach has been used by Höganäs in Sweden and Magnetics International, Inc., in the USA [12.2] [12.3] to develop processes in which the warm pressing of composites produces a solid able to be machined to a finally desired shape. With appropriate design of the machine requiring a core, pressings can be produced to final shape and size.

The rise in demand for motors able to operate at higher speeds via higher (variable) frequency supplies provided by inverters is a challenge to conventional electrical steel strip. The powder composites offer the possibility of keeping eddy losses in check at raised frequencies, combined with the advantage of moulded shapes.

The frequency range of most interest for applications of composites is 100–1000 Hz. The permeability of the pressed material is considerably lowered by the twin effects of the reluctance of the interparticle coating and the consequential demagnetising effects. However where acicular particles are used these effects may be less. Above 1.5 T permeabilities are much reduced compared with strip steel, but at the elevated frequencies for which these materials are designed, solid steels would not be used much above 1.0 T in any case.

Overall core densities close to those attainable in a silicon-steel-laminated core can be obtained. Eddy currents can be further reduced by the alloying of the metal used in the particles with silicon or phosphorus to further reduce eddy losses by raising resistivity. The lack of an easy-to-move domain wall structure will adversely affect the hysteresis losses of composites.

Overall it appears that this class of product will find an expanding niche application in machines where frequencies in the 500–1000 Hz range are being used. The near-net shape pressing possibilities and the ability of flux to flow equally in three dimensions can be expected to give the machine designer extra scope for innovation.

It seems that water atomisation is used to produce powder but that the naturally occurring surface oxide may be supplemented by the insulating effects of the pressing binder. A compromise has to be reached between the beneficial and detrimental effects of more of less per cent binder being used. So far the means for producing acicular particles is a commercial secret, though speculation about chemically grown whiskers has been heard.

12.6   Air via superconductivity

It may seem strange to consider air as a medium for delivering magnetic flux in power devices. However the availability of superconductors enables very large currents to be contained in the turns of superconducting solenoids. This technique is already in use for some varieties of medical scanner where multi-tesla fields are required.

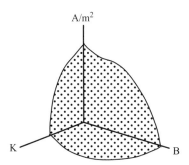

Figure 12.3   The 'corner' of superconductivity. Temperature in kelvins, ambient field in tesla (B), current density in amps/metre^2

Originally liquid helium temperatures (4.2 K) were needed to induce superconductivity, but the progressive development of new materials has led to the production of superconductors able to operate at liquid nitrogen temperatures (77 K). More recently superconductors with higher and higher critical temperatures are being found so that there is a prospect that one day a room-temperature superconductor could be found.

It should not be thought that superconductors enable field strengths to be obtained without limit. Superconducting materials can only tolerate a limited upper current density before reversion to the 'normal' state, and likewise can only tolerate an upper limit of magnetic field. This response can be indicated in principle by Figure 12.3. Here the superconducting region is shown shaded. As temperature is raised this shaded area shrinks into the corner and disappears at T_{crit}.

Despite these limitations there is real prospect that in years to come some class of devices may be created which will enable convenient use of superconductors to provide machines without iron cores. Even so, the free 2 T provided by saturated iron may still be welcome as a supplement to the efforts of superconducting coils.

References

12.1. MOSES, A. J.: 'Electrical steels: past, present and future developments', *Proc. IEE*, 1990, **137**, Pt A (5), pp. 233–45.

12.2. *Soft magnetic composite update*, Vol 1, No 1, 1997. Höganäs AB, S263 83 Höganäs, Sweden.

12.3. Magnetics Int Inc. 3001 East Columbus Drive, East Chicago, Indiana 46312, USA.

Chapter 13

Thickness assessment

Thickness is a vital property of electrical steels. Reducing thickness restrains eddy current loss, but decreasing thickness is expensive and tends to deteriorate the space occupancy of iron. Power loss is assessed at specified peak operating inductions, e.g. 1.5 tesla, and for this quantity to be known the active cross-sectional area of metal is required. Width is comparatively easy to measure, as is length. Loss is quoted as watts/kg and the mass is available from the product of width, length, density and thickness. Ordinarily loss is quoted in watts/kg so that when mass is required, it may be determined directly from a weighing machine. Then thickness (to give access to cross-section) is calculated by use of a conventional figure for density. On other occasions (such as on a production line), thickness may be determined and mass (for watts/kg) calculated from width, length and conventional density.

It might be considered that since space is taken up in the cores of machines by the volume of steel present, loss in watts/m^3 would be more appropriate. However, so entrenched is the commercial act of buying and selling by weight for steel products that such a simplification has little chance of coming into use.

Finished thickness is, of course, the most important and lies in the range 0.2–1.0 mm.

13.1 Magnetic testing

Epstein test samples of known length and width are normally cut by carefully calibrated shears or punch and die methods. Their dimensions may be verified by traceable Vernier gauge to fine limits. To get the cross-sectional area to enable $\hat{B}$ setting, the strips are weighed. Balances of high accuracy are again available with good traceability:

$$\text{Cross-section} = \frac{\text{mass}}{\text{length} \times \text{density}}$$

Density may be a conventionally agreed figure (e.g. 7.65 g/cm^3 for 3% silicon steel). If density is to be determined for a new grade or an out of the ordinary alloy two

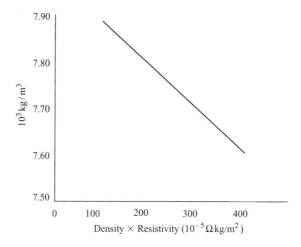

Figure 13.1   Relationship between density and the product of density and resisitivity for steels with silicon and aluminium contents up to 3%. The value of density × resistivity is conveniently derived from measurements on an Epstein strip. Resistance between potential contacts when used as a four-terminal resistor. R = resistivity × electric length/width × thickness, and mass = length × breadth × thickness × density, so that resistivity × density = resistance × mass/length × electric length. This allows resistivity × density to be readily found and density read off from the graph. See also the description in BS6404 Pt. 13, 1996.

methods are possible:

(1)   The relationship between alloy content and internationally accepted density can be used. See Figure 13.1.
(2)   Density can be determined by a weighing method. This involves weighing samples both in air and immersed in water.

With chemical balances able to operate to very close (and traceable) tolerances this sounds simple. In fact the procedure is tedious and difficult.

Account has to be taken of:

(a)   Control of temperature so that the density of the water is known
(b)   Removal of dissolved gas from water (boiled and cooled)
(c)   Removal of adhering bubbles of air on an immersed sample (careful brushing)
(d)   Allowances made for the effects of surface tension at the points where a support stirrup dips into the water. This may be calculated given the temperature and size of the stirrup rod. Alternatively a surface tension reducing agent may be added to water, but at such low levels that the density of water is not noticeably affected.
(e)   A range of representative samples should be measured and the results regressionally compared with figures for alloy content taken from chemical analysis.

The complexity of the procedure makes it clear why conventional densities are much in use. Overall, if density can be reliably determined to ±0.1% it is a good achievement.

Finally the existence of coatings on the sample should either be allowed for by calculation, or these should be removed before measurement is started.

It appears then that in the laboratory access to mass via an accurate balance, and to density via a conventional figure, will allow loss in watts/kg at a known $\hat{B}$ setting to be satisfactorily achieved.

In passing, mention may be made of the stacking factor or space occupancy. This is the percentage of space occupied by metal in a given stack of laminations. A pile of 24 Epstein strips can have its notional thickness (24 × individual) calculated, then be placed in a squeezing jig which exerts a compression on the stack of an intensity typical of machine use. The physical height of the stack (usually assessed by a micrometer dial gauge) is compared with the notional value derived from weighing, etc:

$$\frac{\text{Notional value}}{\text{Dial gauge reading}} \times 100 = \text{stacking factor}$$

Figures in the high 90s are usual. The figure for the stacking factor found will be influenced by the presence of coatings (not much) and by the surface roughness of steel. This latter may have been purposely raised to help avoid sticking and aid gas penetration in a lamination stack, or may be as smooth as practicable.

Within the production plant thickness is seen in various ways. The factors to be considered are:

(a) Need to compute cross-section for on-line magnetic measurement ($\hat{B}$ setting)
(b) Need to verify compliance with intended supply thickness (for market)
(c) Need to control the operation of rolling mills.

In all of these rapid continuous on-line assessment is needed.

The question could be asked, 'What is thickness?'. Theoretically the answer should be 'The distance between the top and bottom surfaces of a piece of sheet metal'. How true, but of course, it begs the question of where the surfaces are deemed to be. Traditionally the engineer would say, I shall apply my micrometer to the steel and give you a figure. The careful engineer would also be using a micrometer traceable to standard slip gauges at a known temperature, and he would take multiple readings without undue force and apply appropriate statistics to the data. But look at Figure 13.2. This shows a notional view of micrometer jaws coming together on a piece of steel. The surfaces are shown (exaggeratedly) as having a finite roughness. The micrometer jaws come to rest on the peaks of the asperities forming the surface. If the thickness were calculated from the weight, the asperities and valleys would be averaged and the surface positions move to X and Y. There will then be a constitutional difference between thickness assessed by micrometer and by weighing.

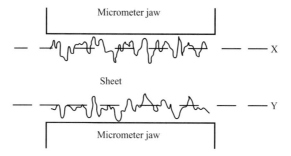

Figure 13.2   Where is a surface defined as existing? Micrometer thickness is the distance between the top and bottom surfaces of a sheet

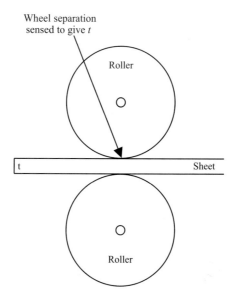

Figure 13.3   Contact micrometer

13.2   On-line measurements

The steel industry has long needed continuous on-line thickness measurements. Various devices have been developed involving contact with moving strip. The sensing components may be hardened steel wheels, Figure 13.3, or balls.

The use of radiation absorption has proved to be a reliable method of thickness assessment. Figure 13.4 shows the basic system. The radiation used has been variously β-radiation from strontium 90, γ-radiation from americium 241 or X-rays generated in an X-ray tube.

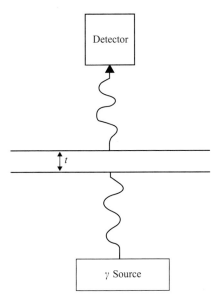

Figure 13.4   Basic scheme for a gamma gauge

13.3   Beta rays

Beta-particles are not very penetrative in steel and particles tend to scatter on encountering the metal surface making effective shielding expensive. Absorption is related to the logarithm of thickness. Alloy additions to steel affect radiation opacity (to β-rays) more rapidly than they affect true density and this complicates allowances being made for different alloy contents.

The quantum statistics of beta-sources of convenient size are such that unless long time constants of detected signal integration are used the thickness figure is uncertain. Long integration times mean that spatial discrimination of thickness change of the strip on the production line is poor.

13.4   X-rays

X-rays can be very penetrative and photon flux levels able to give statistically reliable results very quickly (milliseconds) are easily obtained. X-rays are easily collimated and when the X-ray tube is switched off the device is radiation safe at once. However, X-ray systems use high voltages, circa 100 kV, and considerations of safety and insulation demand expensive arrangements. Figure 13.5 shows an industrial X-ray tube.

The flux of X-radiation is controlled by the cathode emission in the tube and this is rather difficult to control precisely to give a stable and reproducible photon flux. By contrast radio isotopes of many years half-life can be deemed to be absolutely constant on a month-to-month basis.

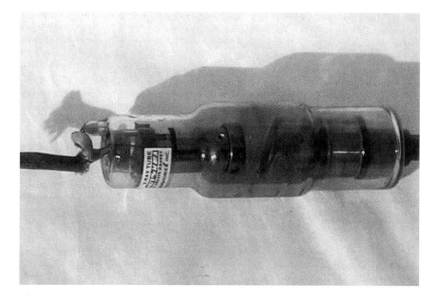

Figure 13.5   Industrial X-ray tube

13.5   Americium 241

Americium 241 is an artificial isotope which produces γ-rays at 60 keV. This is very convenient for measurement of steel in the thickness range 0.2–2 mm where accurate measurement is most needed in electrical steels. It has a half-life of 458 years so that month-to-month photon flux levels are sensibly constant. However americium sources of high photon flux output from a small area are difficult to obtain. This is because the isotope is dense and absorbs its own γ-flux. Making a source plaque thicker does not increase γ-output very much. Making it of greater surface area yields more radiation but the production of a small collimated γ-beam is then more difficult.

A compromise tends to be reached in which beam diameters of some 2–3 cm are used (which gives good spatial resolution of thickness change) with adequate photon statistics such that signal integration over 0.25 second gives a result accurate to about ±0.25%.

Since radiation is absorbed as the log of thickness the 'averaging' of asperities is controlled by different statistics to averaging by weight.

These factors require that a policy of thickness measurement must be developed and adhered to so that all parties are clear as to what is being measured by what method. The policy adopted by European Electrical Steels is that thickness will be measured by radiation gauge and the results expressed in terms of the thickness (as calculated by weighing) of low-carbon mild steel which would give the same radiation absorption.

To calibrate a gauging system on this basis requires access to a system of metal master standards by comparison with which routine standards can be produced.

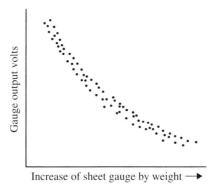

Figure 13.6   Data (notional) scatter on a plot of gauge output volts versus sample thickness

To provide this set of master standards a steel was chosen of known but very low alloy content and a coil of this steel was rolled down to progressively thinner thicknesses covering the range of values relevant to the industry.

A large and expensive operation was mounted to determine the density of this steel using many gravimetric assessments conducted on samples drawn from various positions along a master coil of steel. Similarly its chemical composition and metallurgical cleanliness were extensively checked.

The series of samples produced was then offered to a radiation gauge. The gauge was operated in such a manner that no 'measurement' was attempted, the voltage output from its detector was merely recorded for each offered sample. Each sample had accorded to it a 'thickness by weight' based on the final agreed figure for density, and its superficial dimensions as found from a traceable Vernier gauge.

Separate tests had shown that the gauge detector gave a smooth monotone variation of output as an interposed absorber varied (e.g. slowly filling or emptying a bath of water traversed by radiation). A plot of gauge detector volts versus 'calculated thickness by weight' gave a plot of the form shown in Figure 13.6.

13.6   Group vote

The plot in Figure 13.6 shows a degree of scatter. This will relate to the ability with which the gauge detector output can be exactly read (a long signal integration time is used for this, e.g. 10 seconds) and the degree to which computed sample thicknesses are in error.

The next procedure allocates 'new' values of thickness to each sample such as to bring it exactly onto the mean line through the data of Figure 13.6. A computer plot of this line is readily produced from the data. Clearly the 'new' values of thickness amount to a group vote as to their values.

The samples were re-run through the gauge and a fresh set of readings taken. The 'new' thickness values are then plotted against detector output voltage. This second plot will be much tighter than the last. A further re-run could be used if necessary. At the end of the operation the samples concerned, maybe 200 of them, are split into three lots. One set is titled 'reserve master standards' and safely stored, another set is titled 'transfer standards' and is used to create working standards as needed. The third set is named 'reserve standards'.

As an act of policy the radiation opacity of mild steel as it relates to gauge by weight, dimension and density is the basis of all gauging work in European Electrical Steels.

13.7   Stainless standards

It is recognised that mild steel is a vulnerable metal and liable to corrosion. For this reason stainless steel working standards are used. The specific radiation opacity of this steel is not determined but standards are given 'mild steel equivalent' values by comparison with the master set.

Engineers adjusting and maintaining thickness gauges in the production environment need sets of standards to apply in routine work. To meet this need sets of issued routine-use standards are produced and issued to engineers. These can be recalled for check or recalculated if damaged. Figure 13.7 shows a typical production gauge and Figure 13.8 a set of issue standards.

Just as important as standards is the means used to offer them to gauges being calibrated. Unless a sample intercepts a radiation beam at exactly 90° its effective

Figure 13.7   Radiation gauge on a steel production line

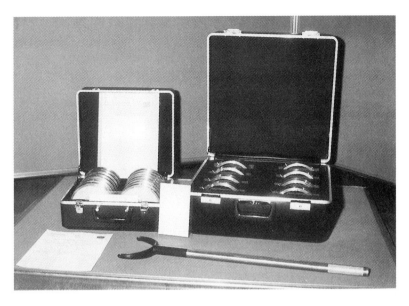

Figure 13.8   Sets of radiation thickness standards

opacity increases, and so does photon scatter. Further, the apparent thickness can vary with the pass line at which the sample is offered. It is important that the pass line of strip to be measured runs at exactly 90° to the radiation beam and the test samples are reliably and reproducibly applied at this height. Consequently purpose-made precision jigs must be provided so that samples are not only accurate in themselves, but are placed into use in precisely controlled positions.

When a sample is being observed by an on-line gauge the signal integration time will be increased to maybe 10 seconds so that the very best statistical appreciation of photon flux is obtained. In actual use the gauge response time constant is shortened and the worsened signal/noise ratio is accepted.

Even though the moment to moment indicated values of thickness may be uncertain to ±0.25% (maybe 0.1% on long integration) the accuracy of the whole system is still high as the integral of errors due to photon uncertainty sums to zero over the length of a coil of steel.

Some commercial thickness gauges incorporate built-in calibration standards. These are often configured so that samples of various thicknesses are called into use simultaneously so that their combined opacities add up to a desired standard value. A system of work based on subdivision of opacities in this way can be used, but it gives a constitutionally different result to that obtained when all full-solid standards of the desired thickness are used. Tests show that several standards operated in series experience inter-sample scatter of photons and must of necessity operate each at different pass lines. European Electrical Steels operates the technique of using all-solid samples to plot the response of a gauge in the range of interest [13.1].

13.8   Alloy compensation

The addition of alloying elements (notably silicon) alters the apparent thickness of material to radiation. Consequently an extensive programme of measurement has been undertaken to produce factors relating apparent thickness to real thickness in the face of varying alloy content. Instead of applying alloy correction to gauges it is found convenient and less confusing to allocate rolling 'aim' thickness in terms of 'mild steel equivalent' so that when steel is rolled (and later assessed) it is its mild steel equivalent that is used in the knowledge that for that alloy this will lead to the desired physical thickness because the mild steel value of that thickness of alloyed material has been pre-computed.

Consumers of electrical steel wish to operate in terms of 'micrometer thickness' so that when mild steel equivalent aim thicknesses are decided a factor must be included to allow for the average expected deviation between the methods used in production and those used by machine builders.

It is clear then that thickness is considered to be a very important parameter for electrical steels. Getting it reliably right controls not only the thickness as it will be seen by users but also the accuracy of the $\hat{B}$ setting in magnetic measuring devices.

Conventional densities are used to decide the mild steel equivalent of steels in production and conventional densities are also used for in-laboratory power loss tests. European Electrical Steels provides standards for themselves and other steel finishing plants in the UK.

An often encountered question is 'How thick is it?' to which the reply may be 'It depends on what you mean by thickness'. The ability to obtain figures which are accurate and traceable is a very valuable facility. A side effect of access to precision standards is the ability to exploit the full capability of modern gauges to deliver exact results.

13.9   Further trends

While radiation thickness gauges are a well-developed part of the steel industry, it is true that the presence of high voltages (X) and long-life radio isotopes (γ) are features which have some associated hazards. To address this problem, efforts have been made to use other techniques to infer thickness.

13.10   Resistance methods

It has been found possible to use the 1m wide steel strip in a production line as a four-terminal resistor. If an electric current is introduced into strip and removed 2 or 3 metres further along, the potential drop along the strip (with known current) allows the ohms/metre of strip to be noted.

For moving strip, edge electrodes can be used to inject several ampères into strip. Appropriate volt drops can be measured with subsidiary sliding potential electrodes. Ref. [13.2] describes this technique. Current is kept constant by feeding it from a high-impedance source so that contact volt drops do not affect the current level. The use of conventional resistivities enables the strip cross-section to be obtained and can control the $\hat{B}$ setting directly. Knowledge of width makes thickness itself available for other quality-control purposes.

13.11   Magnetic weighing

Recent research has shown that when steel is very heavily saturated (magnetically) the flux produced relates only to the cross-sectional area of the metal and its alloy content. This means that by reversing magnetisation from positive saturation to negative saturation and noting the flux change the cross-section of the metal can be determined.

This subject has been deeply researched [13.3] and the findings are that by allocating to steel types a 'conventional saturation polarisation – J_{sat}' the magnetic saturation technique can be used to provide very accurate values for cross-sectional area. Such measurement involves excitation at 50 Hz only and there is no requirement for waveform control, so equipment is based on simple electrical engineering. Magnetic weighing of this sort can be viewed in various ways:

(1)   Used to 'weigh' the steel in an Epstein frame while it is there for power loss testing. This would avoid the need for traditional laboratory mass determination.
(2)   Used on a production line it could provide an accurate cross-sectional area signal for on-line $\hat{B}$ setting in an on-line loss tester. Combined with width information thickness is available for ordinary production control purposes. Speed of response can be rapid, e.g. 1/10 second.

A great deal of work was needed to show the true accuracy of magnetic weighing. The required applied fields, e.g. 70 kA/m, had to be determined in the light of prevailing demagnetising factors. Conventional J_{sat} values had to be determined for a range of material types.

The constitutional difference in the determination of thickness (thus cross-section) by gravimetric and radiation methods had to be closely examined by studying a very large number of very carefully dimensioned samples by the three methods. The results show that differences between the three methods become lost in the 'noise' associated with the extremes of precision on any one.

So, magnetic weighing can be used with no loss of traceable accuracy. Ref. [13.3] gives a useful account of this work.

13.12   Profile

If cross-sectional area is determined by magnetic weighing then only the average cross-section is found. If cross-section is determined from radiation thickness at any one place across the strip width the computed cross-section will depend on the assumption that thickness at that point is a good representation of its value across the width. Gauges which traverse the whole width of the strip repeatedly are available (profile gauges) and are useful for finding unwelcome levels of centre 'crown'. However when the whole sheet is magnetised any variation of thickness will lead to differing values related to how flux has disposed itself with respect to strip width. $\hat{B}$ values higher or lower than the mean lead to losses above or below the mean in those locations.

Thickness therefore is a vital parameter to be traceably measured and controlled at every phase of the production of electrical steel.

References

13.1. BECKLEY, P.: 'Master standards for radiation thickness gauging', Sheet Metal Industries, September 1974, pp. 598–601.
13.2. ARIKAT, M., BECKLEY, P. and MEYDAN, T.: 'A novel cross sectional area sensor for on-line power loss determination in electrical steels', *IIT conference on Magnetic materials*, Chicago, May 1997.
13.3. BECKLEY, P., CAO, J. Z. and SHIRKHOOI, G. H.: 'Cross section, thickness and the approach to ferromagnetic saturation', *IIT conference on Magnetic materials*, Chicago, May 1998.

Chapter 14

Digest of standards

Familiarity with National and International Standards is a very valuable resource. When discussions of quality, characteristics and test methods arise the ability to refer easily to the relevant standards saves a lot of repetitive argument. It is usually easiest to discuss how features of a material diverge from a standard than to start from a clean sheet. Standards are revised and updated from time to time and it is advisable to maintain contact with standards bodies so that the standards one may be using are the current ones.

The purchase of a full set of British Standards listed as covering most aspects of electrical steels is an expensive undertaking (some hundreds of pounds in 1999, for 18 documents), but the notes given hereafter will guide the reader as to which if not all standards are relevant to his operation. Parties using exclusively grain-oriented steel would not wish to concern themselves with non-oriented specifications, etc. However a familiarity with a wide range of test methods and procedures is worth acquiring and a set of current standards held in one's library is a valuable asset. In this chapter a list is given of the key British Standards which apply and the address of The British Standards Institution.

Of course foreign standards are encountered from time to time, but more and more alignment between British Standards, IEC (International Electrotechnical Commission) and European (Euronorm) standards is becoming complete.

The ASTM (American Society for Testing Materials) standards used in the United States of America remain committed to some of their own procedures, but it may be hoped that world standardisation can become usefully complete in the foreseeable future.

Some time ago British Standard BS601 dealt with Electrical Steels, and copies may still be found, however the progressive harmonisation of national and international standards has overtaken earlier domestic standards. Different standards evolve and are reviewed over varying time cycles, so that when requesting a standard it is useful to enquire about its state of revision. A standard may carry an origination date of some years ago and not be in need of updating, whereas a more recent one may be in the throes of revision. Because of progressive dual numbering of standards it is sufficient to set out here a British Standard list which will in fact represent the key range of standards worldwide.

14.1   The British Standard BS 6404 Part 1 (1986) – Magnetic materials classification

This is a classification of many sorts of magnetic materials, and covers electrical steels.

Useful pointers are:

(1)   *'Soft' magnetic materials* have a coercive force of less than 1000 A/m, and for 'hard' materials it is greater than 1000 A/m.
(2)   *Soft irons* normally have carbon levels below 0.03% and are employed in DC devices.
(3)   *Low-carbon mild steels.* Usually classified in terms of power loss at $B_{max} = 1.5$ T and 50/60 Hz. These are supplied in sheet form in the range 0.47–1.0 mm. They are used in the laminated cores of small electrical machines. Power losses at $B_{max} = 1.5$ T, 50 Hz, 0.65 mm thick range up to 12 W/kg.
(4)   *Silicon steels.* These are available in solid bar form with silicon ranging up to 5% Si. The materials resistivity is linked to %Si and is around 65 microhms cm at 5%. Application is in clutches, relays, etc. Flat sheet silicon steel is classified in terms of power loss at $B_{max} = 1.5$ T, 50 Hz. Thickness normally lies in the range 0.35–0.65 mm. Power losses at $B_{max} = 1.5$ T, 50 Hz range from 2.0 to 10 W/kg. The principal application is in the cores of rotating machines.
(5)   *Grain oriented steels.* These are anisotropic steels largely used for transformer cores. Typical thicknesses are 0.27, 0.30 and 0.35 mm, though material is available at 0.23 mm and 0.5 mm. A typical power loss at $B_{max} = 1.5$ T for a thickness of 0.27 mm may be 0.8 W/kg.
(6)   *High-strength magnetic steel sheet.* This material has UTS values in the range 300–700 N/mm^2 and proof stress in the range 150–525 N/mm^2. This material is used where strength has to be combined with adequate magnetic permeability.

BS6404 details a wealth of other information relating to the steels mentioned as well as to nickel irons and cobalt iron steels.

An IEC Classification Document CEI/IEC 6404-8-6, 1999 is also now available.

14.2   British Standard BSEN.10126 (1996) – Cold rolled electrical non-alloyed steel sheet and strip delivered in the semi-processed state

This is the standard for silicon-free motor lamination steel. The final anneal to be given before test involves a decarburising anneal of 2 hours at 790 °C. The atmosphere is to be 20% hydrogen, 80% nitrogen at a dew point of +25 °C.

14.3   British Standard BS EN 10165 1996 – Cold rolled electrical alloyed steel sheet and strip delivered in the semi-processed state

This is the product standard for cold rolled semi-processed electrical steel strip containing significant amounts of alloying elements, and which will require heat treatment to develop fully its magnetic properties.

This standard deals with the tolerances allowed on thickness, edge camber, flatness and outlines the main magnetic characteristics. These apply after an anneal at the reference temperature in hydrogen containing 20% by volume of H_2O (dewpoint 20 °C).

Conventional density range kg/dm³	Temperature of final anneal
7.65 ⎱ 7.70 ⎰	840 ± 10 °C
7.75 ⎱ 7.80 ⎰	790 ± 10 °C

This standard contains a detailed set of conditions relating to supply, packaging, etc.

14.4   British Standard BS 6404 Section 8.4 (1986) – Specification for cold rolled, non-oriented magnetic steel sheet and strip delivered in the fully annealed state

[This standard has now been replaced by BSEN 10106, (1996).] This is the product standard for non-oriented fully annealed strip which may be used without further heat treatment. The standard deals with magnetic properties, the tolerances allowed on thickness, edge camber, flatness and residual curvature, as well as with stacking factor, ductility, internal stress and characteristics of coatings.

Magnetic ageing of the steel can be assessed by re-testing after heating for 24 hours at 225 °C or 600 hours at 100 °C. Anisotropy of magnetic properties can be assessed by making separate measurements in the directions parallel and transverse to the rolling direction.

14.5   British Standard BS EN 10107 (1996) – Grain-oriented electrical steel sheet and strip delivered in the fully processed state

This is the product standard for grain-oriented electrical steels having strong directional (anisotropic) properties. The standard deals with magnetic properties and mechanical features such as edge camber, flatness, ductility and internal stress and surface coating resistance and burr height.

14.6 British Standard BS EN 10265 (1996) – Steel sheet and strip with specified mechanical properties and permeability

This standard deals with magnetic materials, with specific mechanical properties and magnetic permeability such as may be required to support high mechanical loads.

Details of the form of supply and mechanical tolerances are given. Hot rolled materials lie in the range 1.6–4.5 mm thick and cold rolled 0.5–2.0 mm thick. The proof stress (0.2%) values for hot rolled steel lie in bands from $250\,\text{N}/\text{mm}^2$ up to $700\,\text{N}/\text{mm}^2$. These values are associated with flux densities at 15 kA/m between 1.80 and 1.78 T. Cold rolled grades can provide 1.8 T at 15 kA/m for the $400\,\text{N}/\text{m}^2$ proof stress.

The magnetic properties of the material are measured using the Epstein frame method.

14.7 British Standard BS EN 60404 – 2 (1998) – Methods of measurement of the magnetic properties of electrical steel sheet and strip by means of an Epstein frame

This standard details the procedures for the Epstein test method. Guidance is given as to winding formats, which are useful. 700 or 1000 turns are usual for both primary and secondary windings. A power supply with a frequency stability of 0.2% is needed and arrangements to retain a form factor of the secondary voltage of $1.111 \pm 1\%$.

Full details are given for the procedures for the measurement of power loss, specific apparent power and permeability. Any attempt to set up and operate an Epstein frame requires careful study of this standard (or its ASTM equivalent) or the procurement of commercial equipment built in accordance with this standard and verified via a NAMAS laboratory.

14.8 British Standard BS 6404 Pt 3 (1992) – Methods of measurement of the magnetic properties of magnetic sheet and strip by means of a single sheet tester

This standard details the construction and operation of single sheet testers. In particular the flux closure yokes and their construction are specified along with the nature and dimensions of windings.

Apart from considerations of path length, the operation of the tester is much like that of an Epstein frame. Guidance is given about how S.S.T. calibration should or should not be related to Epstein values (Figure 14.1).

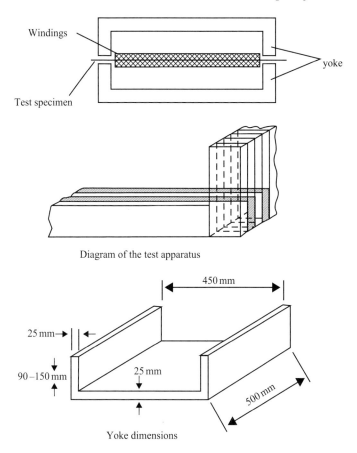

Diagram of the test apparatus

Yoke dimensions

Figure 14.1   Single sheet tester arrangement (courtesey BSI)

14.9   British Standard BS EN 60404 – 4 (1997) – Methods of measurement of DC magnetic properties of iron and steel

This standard deals with measurement of the DC properties of iron and steel by the permeameter and other methods.

Ring method. Advice is given about ring dimensions, how to form the ring and how to wind it. Problems of overheating are considered.

The permeameter method. In this method a sample is enclosed in measuring windings and the flux path completed by massive low reluctance steel yokes. Exact instruction is given for the measurement of steel properties and the completion of BH loops. Figure 14.2 shows one realisation of the permeameter.

This standard gives a very good view of the scope of permeability measurements and how they should be made if they are to be the basis of commercial activity.

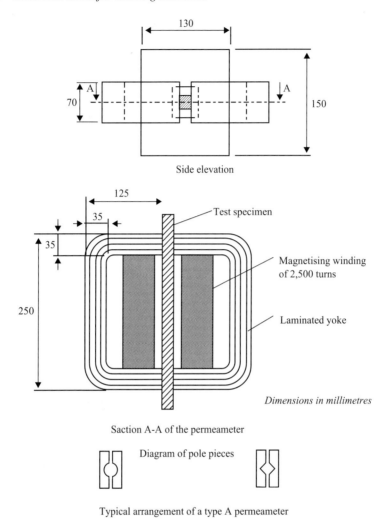

Figure 14.2   Diagram of a permeameter (courtesey BSI)

14.10   British Standard BS 6404 Part 11 (1991) – Method of test for the determination of the surface insulation resistance of magnetic sheet and strip

This standard deals with methods of test for the surface insulation of electrical steel sheet. A description of apparatus and the appropriate statistics to be applied to test results are given. See also BS 6404 Part 20 1996 – Resistance and temperature classification of insulation coatings.

14.11 British Standard BS 6404 Part 12 (1993) – Guide to methods of assessment of temperature capability of interlaminar insulation coatings

This standard gives guidance on the assessment of the temperature resist capability of interlaminar insulation coatings. This is a type test used to characterise types of insulant and would not be expected to be applied frequently.

14.12 British Standard BS 6404 Part 13 1996 – Methods of measurement of density, resistivity and the stacking factor of electrical sheet and strip

This standard deals with methods of measurement of density and stacking factor of electrical steels. Particularly a method for determining density through the relationship between density and alloy content is described.

References 1, 2 and 3 within the standard merit attention. These are:

[1]   SCHMIDT, K. H. and HUNEUS, H.: 'Determination of the density of electrical steel made from iron-silicon alloys with small aluminium content'. *Techn. Messen*, **48**, 1981, pp. 375–379.
[2]   VAN DER PAUW, L. J.: 'A method of measuring specific resistivity and Hall effect of discs of arbitrary shape'. *Philips Research Reports*, **13**, 1958, pp. 1–9.
[3]   SIEVERT, J.: 'The determination of the density of magnetic sheet steel using strip and sheet samples'. *J. Magn. Mater.*, **133**, 1994, pp. 390–392.

14.13 British Standard BS 6404 Part 20 (1996) – Resistance and temperature classification of insulation coatings

This document sets out a single electrode test for surface insulation assessment including the appropriate statistics to be used on test results. Reference is also made to time–temperature performance of coatings. Although primarily descriptive of temperature classification of coatings, this standard contains a full account of the UK single-electrode method. This may be valuable for its relationship to some UK specifications for steel.

14.14 British Standard BS EN 10251 (1997) – Methods of determination of the geometrical characteristics of electrical sheet and strip

This standard details methods for the determination of flatness, residual curvature, edge camber, internal stress and burr height.

14.15   British Standard BS 6404 section 8.8 (1992) – Specification for thin magnetic steel strip for use at medium frequencies

This is the specification document for thin magnetic strip for use at medium frequencies. Thicknesses from 0.05 mm to 0.15 mm are covered. Frequencies of 400–1000 Hz are normally relevant at peak inductions of 1.0 or 1.5 T.

14.16   British Standard BS EN 10252 (1997) – Methods of measurement of magnetic properties of magnetic steel sheet and strip at medium frequencies

Design advice is given for the construction of Epstein frames for use up to 10 kHz. The measurement system is based on the wattmeter method for power loss assessment. In former times bridge methods have been included in the field of measurement, but are now discontinued. If insight into bridge methods is sought then the now obsolete BS 601 may be consulted. Also dealt with in: B. Hague '*Alternating current bridge methods*', Pitman 1938, and later editions.

14.17   British Standard BS 6404 section 8.10 (1994) – Specification for magnetic materials (iron and steel) for use in relays

This standard is the specification document for iron and steel used in relays and like devices. Physical properties are described. The primary characterisation of material refers to coercive force with grades in the region 40–240 A/m. This applies after a 790 °C anneal in 20% hydrogen, 80% nitrogen, water being added to achieve a dewpoint of 35 °C.

14.18   British Standard BS 6404 Part 7 (1986) – Method of measurement of the coercivity of magnetic materials in an open magnetic circuit

This standard describes methods of measuring the coercive force of steel in an open magnetic circuit. This is a rapid and convenient method of determining the coercive force of steels. Samples can be of widely varying shape and measurement is rapid. The test method relates mainly to relay steels, but is of general usefulness.

14.19   General – addresses

(1)   British Standards Institution
389 Chiswick High Road
London W4 4AL UK.

(2)   International Electrotechnical Commission
3 Rue de Varembé
Geneva
Switzerland.

(3)   American Society for Testing Materials (ASTM)
100 Barr Harbor Drive
West Conshohocken
Pennsylvania, 19428-2959
USA.

Chapter 15

Data and curves

This section contains a selection of material property graphs and tables which will be of use to the machine designer. These have been made available by the kind permission of European Electrical Steels (now Cogent Power Ltd.)

All data given is accurate to the best of the provider's knowledge and are given only to help the reader make his/her own evaluations and decisions. No liability is accepted by the providers for the consequential results of usage and no quality guarantees are implied.

15.1   Non-oriented fully annealed steels

Table 15.1   Typical magnetic properties at 50 Hz for fully processed electric steel

Grade EN 10106	Specific total loss at 50 Hz		Anisotropy of loss	Magnetic polarisation at 50 Hz			Coercivity (DC)	Relative permeability at 1.5 T
	$\hat{J} = 1.5\,\text{T}$ W/kg	1.0 T W/kg	%	$\hat{H} = 2500$ T	5000 T	10000 A/m T	A/M	
M235-35A	2.25	0.92	10	1.53	1.64	1.76	35	660
M250-35A	2.35	0.98	10	1.53	1.64	1.76	40	630
M270-35A	2.47	1.01	10	1.54	1.65	1.77	40	730
M300-35A	2.62	1.10	10	1.55	1.65	1.78	45	810
M330-35A	2.93	1.18	10	1.56	1.66	1.78	45	830
M250-50A	2.38	1.02	10	1.56	1.65	1.78	30	800
M270-50A	2.52	1.07	10	1.56	1.65	1.78	30	830
M290-50A	2.62	1.14	10	1.56	1.65	1.78	35	800
M310-50A	2.83	1.23	10	1.57	1.66	1.79	40	930
M330-50A	3.03	1.29	10	1.57	1.66	1.79	40	950
M350-50A	3.14	1.33	9	1.58	1.67	1.79	45	960
M400-50A	3.58	1.54	9	1.58	1.67	1.79	50	1020
M470-50A	4.05	1.79	6	1.59	1.68	1.80	60	1120
M530-50A	4.42	1.96	6	1.59	1.68	1.80	70	1150
M600-50A	5.30	2.39	6	1.63	1.72	1.83	85	1620
M700-50A	6.00	2.72	5	1.64	1.72	1.84	100	1680
M800-50A	7.10	3.22	5	1.65	1.73	1.85	100	1680
M940-50A	8.10	3.68	5	1.65	1.73	1.85	110	1660
M330-65A	3.15	1.35	8	1.57	1.66	1.78	40	910
M350-65A	3.23	1.41	8	1.57	1.67	1.78	40	930
M400-65A	3.63	1.57	7	1.58	1.67	1.79	45	1050
M470-65A	4.06	1.79	6	1.59	1.68	1.80	50	1130
M530-65A	4.35	1.90	4	1.59	1.68	1.80	60	1150
M600-65A	4.95	2.19	3	1.61	1.70	1.81	70	1300
M700-65A	6.20	2.76	3	1.63	1.72	1.83	85	1570
M800-65A	6.90	3.09	3	1.64	1.73	1.84	100	1590
M1000-65A	8.86	4.01	1	1.65	1.74	1.85	110	1600
M700-100A	6.24	2.83	1	1.59	1.68	1.80	50	970
M800-100A	7.20	3.28	0	1.60	1.69	1.80	70	1030

Ageing

Electrical steel should be as far as possible free from magnetic ageing.

Ageing, or the increase of power loss with time, is caused by too high a carbon content in the steel. The carbon content is monitored by analytical equipment to ensure freedom from ageing. Magnetic test samples are given a rapid ageing treatment by heating at 225 °C $\pm$ 5 °C for a duration of 24 hours, and cooling to ambient temperature before testing, as described in EN 10106.

All typical data refer to aged samples, except those with a thickness of 1.0 mm.

Table 15.2   *Specific total loss at 60 Hz (W/kg and W/lb at $\hat{J} = 1.5\,T$)*

Grade EN 10106	Thickness	Maximum*		Typical	
	mm	W/kg	W/lb	W/kg	W/lb
M235-35A	0.35	2.97	1.35	2.84	1.29
M250-35A	0.35	3.14	1.43	2.97	1.35
M270-35A	0.35	3.36	1.53	3.13	1.42
M300-35A	0.35	3.74	1.70	3.33	1.51
M330-35A	0.35	4.12	1.87	3.70	1.68
M250-50A	0.50	3.21	1.46	3.02	1.37
M270-50A	0.50	3.47	1.58	3.19	1.45
M290-50A	0.50	3.71	1.68	3.33	1.51
M310-50A	0.50	3.95	1.79	3.59	1.63
M330-50A	0.50	4.20	1.91	3.83	1.74
M350-50A	0.50	4.45	2.02	3.97	1.80
M400-50A	0.50	5.10	2.32	4.54	2.06
M470-50A	0.50	5.90	2.68	5.13	2.33
M530-50A	0.50	6.66	3.02	5.59	2.54
M600-50A	0.50	7.53	3.42	6.72	3.05
M700-50A	0.50	8.79	3.99	7.60	3.45
M800-50A	0.50	10.06	4.57	8.99	4.08
M940-50A	0.50	11.84	5.38	10.26	4.66
M330-65A	0.65	4.30	1.95	3.99	1.81
M350-65A	0.65	4.57	2.07	4.09	1.86
M400-65A	0.65	5.20	2.36	4.60	2.09
M470-65A	0.65	6.13	2.78	5.14	2.33
M530-65A	0.65	6.84	3.11	5.51	2.50
M600-65A	0.65	7.71	3.50	6.27	2.84
M700-65A	0.65	8.98	4.08	7.84	3.56
M800-65A	0.65	10.26	4.66	8.74	3.97
M1000-65A	0.65	12.77	5.80	11.21	5.09
M700-100A	1.00	9.38	4.26	7.91	3.59
M800-100A	1.00	10.70	4.86	9.11	4.14

*The values of the maximum loss are given as indicative values.

Table 15.3   Typical physical and mechanical properties

Grade EN 10106	Conventional density	Resistivity	Yield strength	Tensile strength	Young's modulus (E)		Hardness HV5 (VPN)
					RD	TD	
	kg/dm³	μΩ cm	N/mm²	N/mm²	N/mm²	N/mm²	
M235-35A	7.60	59	430	550	185 000	200 000	210
M250-35A	7.60	55	430	550	185 000	200 000	210
M270-35A	7.65	52	390	510	185 000	200 000	190
M300-35A	7.65	50	380	500	185 000	200 000	180
M330-35A	7.65	44	330	470	200 000	220 000	160
M250-50A	7.60	59	440	560	175 000	190 000	220
M270-50A	7.60	55	440	560	175 000	190 000	220
M290-50A	7.60	55	440	560	185 000	200 000	220
M310-50A	7.65	52	390	510	185 000	200 000	190
M330-50A	7.65	50	380	500	185 000	200 000	180
M350-50A	7.65	44	330	470	200 000	210 000	160
M400-50A	7.70	42	320	460	200 000	210 000	150
M470-50A	7.70	39	315	450	200 000	210 000	150
M530-50A	7.70	36	310	440	200 000	210 000	140
M600-50A	7.75	30	300	410	210 000	220 000	125
M700-50A	7.80	25	300	410	210 000	220 000	125
M800-50A	7.80	23	300	410	210 000	220 000	125
M940-50A	7.85	18	300	410	210 000	220 000	125
M330-65A	7.60	55	440	560	185 000	205 000	220
M350-65A	7.60	52	380	500	185 000	205 000	180
M400-65A	7.65	44	330	470	185 000	205 000	160
M470-65A	7.65	42	320	460	185 000	205 000	150
M530-65A	7.70	39	315	450	190 000	210 000	150
M600-65A	7.75	36	310	440	190 000	210 000	140
M700-65A	7.75	30	300	410	210 000	220 000	125
M800-65A	7.80	25	300	410	210 000	220 000	125
M1000-65A	7.80	18	300	410	210 000	220 000	125
M700-100A	7.65	44	325	450	185 000	200 000	150
M800-100A	7.70	39	315	440	185 000	200 000	140

RD represents the rolling direction.
TD represents the transverse direction. Values for yield strength (0.2% proof strength) and tensile strength are given for the rolling direction. Values for the transverse direction are about 5% higher.

Table 15.4   *SURALAC® coatings. The coatings listed here relate to Cogent Power products, other producers will offer other coatings*

Designation	SURALAC 1000	SURALAC 3000	SURALAC 5000	SURALAC 7000
Type	Organic	Organic with fillers	Semi-organic	Inorganic
Description	Organic phenolic resin	Organic synthetic resin with inorganic fillers	Organic resin with phosphates and sulphates	Inorganic phosphate based coating with inorganic fillers and some organic resin
Old SURA designation	C-3	C-6	S-3	C-4 / C-5
AISI type (ASTM A 677)	C-3	C-6[1]	–	C-4 / C-5[2]
Thickness range, per side	0.5–7 μm	3–7 μm	0.5–2 μm	0.5–5 μm
Standard thickness	2.5 μm	6 μm	1.2 μm	2 μm
Number of coated sides	1 or 2	1 or 2	2	2
Colour	Yellow to brown	Grey	Brown to grey	Grey
Temperature capability in air (continuous)	180 °C	180 °C	200 °C	230 °C
Temperature capability in inert gas (intermittent)	450 °C	500 °C	500 °C	850 °C
Withstands				
Stress relief annealing[3]	–	–	–	Yes
Burn-out repair	–	Yes	–	Yes
Aluminium casting	Yes	Yes	Yes	Yes
Chemical resistance				
Stamping lubricants[4]	Yes	Yes	Yes	Yes
Transformer oils	Yes	Yes	Yes	Yes
Freon	Yes	Yes	Yes	Yes
Typical pencil hardness	8–9 H	8–9 H	8–9 H	9 H

Table 15.4   Continued

SURALAC®	1007	1025	1060	3040	3060	5007	5012	7007	7020	7040[5]
Typical thickness, μm per side	0.7	2.5	6	4	6	0.7	1.2	0.7	2	4
Typical welding[6]	good	spec	spec	spec	spec	exc	exc	exc	good	mod
Typical punching[6]	exc	exc	good	good	mod	good	exc	good	good	mod
Surface insulation resistance (Franklin ASTM A 717)										
Typical value, Ω cm² per lamination	5	50	>200	100	>200	5	20	5	50	100
Typical value, ampères per side	0.55	0.11	<0.03	0.06	<0.03	0.55	0.25	0.55	0.11	0.06

Please note that all data are typical, not guaranteed.

[1] C–6 is not an official AISI designation.
[2] Suralac® 7000 is classified as a C–5 coating but it can be used as a C–4 coating.
[3] Stress relief annealing in inert or preferably in slightly oxidising atmosphere.
[4] Testing includes all lubricants used by our present customers. New lubricants may need special consideration.
[5] The last two digits of the designation indicates the nominal coating thickness in 0.1 μm.
[6] exc = excellent, good = good, mod = moderate, spec = special precautions/techniques needed.

Dimensions, ranges and tolerances – The ranges given relate to a specific supplier. Alternatives may be found

Dimensions

Cogent Power electrical steels are supplied as slit coils or cut sheet in the following thicknesses and widths:

Table 15.5

Thickness	Max. width for slit coils and sheets	Maximum sheet length
mm	mm	mm
0.35	1170	3500
0.50	1250	3500
0.65	1250	3500
1.00	1250	3500

Minimum sheet length 400 mm.

*Table 15.6   Coil   width   standard tolerances**

Over	Up to and including	Width tolerance
mm	mm	mm
10	150	0/ + 0.2
150	300	0/ + 0.3
300	600	0/ + 0.5
600	1000	0/ + 1.0
1000	1250	0/ + 1.5

*In accordance with EN 10106 and also fulfilling requirements of IEC 404-8-4.

Table 15.7   Coil   width   special tolerances

Over	Up to and including	Width tolerance
mm	mm	mm
10	300	±0.08
300	600	±0.20
600	1250	±0.30

*Table 15.8   Cut length tolerances**

Over	Up to and including	Length tolerance
mm	mm	mm
400	3500	0/ + 0.5% (max. 6 mm)

*In accordance with EN 10106 and also fulfilling requirements of IEC 404-8-4.

Internal diameter of coil

The internal coil diameter is nominally 508 mm (20 in.).

Maximum coil width

The maximum coil width is 1250 mm.

Table 15.9   Thickness tolerance[1]

Nominal thickness	Max. derivation from nominal thickness	Max. difference in thickness, parallel to rolling direction[2]	Max. difference in thickness, at right angle to rolling direction[3]
mm	%	%	μm
0.35	±8	8	20
0.50	±8	8	20
0.65	±6	6	30
1.00	±6	6	30

Maximum coil weight and OD

The maximum coil weight is 20 tonnes or 20.0 kg per mm coil width. The maximum coil outside diameter is 1850 mm.

Geometric characteristics

Cogent Power electrical steels meet all the requirements on geometric characteristics and tolerances (including edge camber and flatness) specified in the standards EN 10106 and IEC 404-8-4.

[1] In accordance with EN 10106 and also fulfilling requirements of IEC 404-8-4.
[2] Within a sheet or a 2 m length of strip.
[3] Measured at least 30 mm from the edges.

Data and curves 233

Table 15.10   *Comparison of Cogent Power grades and international standards*

Core loss 1.5T 50Hz W/kg	EES grade EN10106 (1995)	Previous SURA grade (1987)	IEC 404-8-4 (1986)	DIN 46400 Teil 1 (1983)	JIS C2552 (1986)	GOST 21427.2 (1983)	ASTM A677 (1996)	Core loss 1.5T 60Hz W/lb	ASTM A677M (1996)	Core loss 1.5T 50Hz W/kg	Old, AISI grade
2.35	M235-35A	(CK-27)			(35A230)						
2.50	M250-35A	CK-30	250-35A5	V250-35A	35A250	2413	36F145	1.45	36F320M	2.53	M-15
2.70	M270-35A	CK-33	270-35A5	V270-35A	35A270	2412	(36F158)	1.58	(36F348M)	2.76	(M-19)
3.00	M300-35A	CK-37	300-35A5	V300-35A	35A300	2411	(36F168)	1.68	(36F370M)	2.93	(M-22)
3.30	M330-35A	CK-40	330-35A5	V330-35A			36F190	1.90	36F419M	3.32	M-36
2.50	M250-50A										
2.70	M270-50A	CK-26	270-50A5	V270-50A	50A270	2414					
2.90	M290-50A	CK-27	290-50A5	V290-50A	50A290	2413	(47F168)	1.68	(47F370M)	2.93	(M-15)
3.10	M310-50A	CK-30	310-50A5	V310-50A	50A310	2412	(47F174)	1.74	(47F384M)	3.04	(M-19)
3.30	M330-50A	CK-33	330-50A5	V330-50A			47F190	1.90	47F419M	3.31	M-27
3.50	M350-50A	CK-37	350-50A5	V350-50A	50A350	(2411)	(47F205)	2.05	(47F452M)	3.57	(M-36)
4.00	M400-50A	CK-40	400-50A5	V400-50A	50A400	2216	47F230	2.30	47F507M	4.01	M-43
4.70	M470-50A	CK-44	470-50A5	V470-50A	50A470	(2214)	(47F280)	2.80	(47F617M)	4.89	(M-45)
5.30	M530-50A	DK-59	530-50A5	V530-50A		(2211)	47F305	3.05	47F672M	5.32	M-47
6.00	M600-50A	DK-66	600-50A5	V600-50A	50A600	2112					
7.00	M700-50A	DK-70	700-50A5	V700-50A	50A700	2111	47F400	4.00	47F882M	6.98	
8.00	M800-50A		800-50A5	V800-50A	50A800	2011	(47F450)	4.50	(47F992M)	7.86	
9.40	M940-50A				(50A1000)						

Table 15.10   Continued

Core loss 1.5 T 50 Hz W/kg	EES grade EN10106 (1995)	Previous SURA grade (1987)	IEC 404-8-4 (1986)	DIN 46400 Teil 1 (1983)	JIS C2552 (1986)	GOST 21427.2 (1983)	ASTM A677 (1996)	Core loss 1.5 T 60 Hz W/lb	ASTM A677M (1996)	Core loss 1.5 T 50 Hz W/kg	Old, AISI grade
3.30	M330-65A			V330-65A							
3.50	M350-65A		350-65A5	V350-65A			(64F208)	2.08	(64F159M)	3.62	(M-19)
4.00	M400-65A	CK-37	400-65A5	V400-65A			(64F225)	2.25	(64F496M)	3.92	(M-27)
4.70	M470-65A	CK-40	470-65A5	V470-55A			64F270	2.70	64F595M	4.70	M-43
5.30	M530-65A	CK-44	530-65A5	V530-65A			(64F320)	3.20	(64F705M)	5.59	(M-45)
6.00	M600-65A	DK-59	600-65A5	V600-65A							
7.00	M700-65A	DK-66	700-65A5	V700-65A			64F400	4.00	64F882M	6.98	
8.00	M800-65A	DK-70	800-65A5	V800-55A							
10.00	M1000-65A		1000-65A5	(V940-65A)			(64F550)	5.50	(64F1212M)	9.60	
7.00	M700-100A	CK-37									
8.00	M800-100A										

Note: A designation within brackets, e.g. (35A230) indicates approximate equivalence.

Conversion factors

1 tesla (T) $= 1$ weber/m^2 (Wb/m^2) $= 10\,000$ gauss $= 64.5$ kilolines/sq.in
1 A/m $= 0.01$ A/cm $= 0.0254$ A/in $= 0.01257$ oersted
1 W/kg $= 0.4536$ W/lb (at the same frequency)
1 VA/kg $= 0.4536$ VA/lb (at the same frequency)
1 N/mm^2 (MPa) $= 145.0$ psi (lbs/sq.in.)

Table 15.11   Typical specific total loss data, W/kg at 50 Hz

Grade EN 10106	Thickness mm	Specific total loss, W/kg at 50 Hz and a magnetic polarisation J (T) of								
		0.90	1.00	1.10	1.20	1.30	1.40	1.50	1.60	1.70
M235-35A	0.35	0.77	0.92	1.10	1.31	1.57	1.91	2.25	2.54	2.75
M250-35A	0.35	0.81	0.98	1.16	1.37	1.66	2.00	2.35	2.66	2.87
M270-35A	0.35	0.84	1.01	1.20	1.42	1.70	2.08	2.47	2.80	3.05
M300-35A	0.35	0.92	1.10	1.30	1.54	1.82	2.21	2.62	2.98	3.25
M330-35A	0.35	0.99	1.18	1.40	1.66	1.99	2.42	2.93	3.46	3.86
M250-50A	0.50	0.85	1.02	1.20	1.41	1.67	2.02	2.38	2.72	2.99
M270-50A	0.50	0.90	1.07	1.27	1.50	1.77	2.13	2.52	2.87	3.14
M290-50A	0.50	0.96	1.14	1.35	1.59	1.88	2.24	2.62	2.94	3.18
M310-50A	0.50	1.03	1.23	1.46	1.71	2.00	2.40	2.83	3.24	3.58
M330-50A	0.50	1.08	1.29	1.53	1.80	2.12	2.54	3.03	3.45	3.78
M350-50A	0.50	1.11	1.33	1.57	1.85	2.18	2.63	3.14	3.65	4.05
M400-50A	0.50	1.28	1.54	1.82	2.14	2.52	3.01	3.58	4.16	4.64
M470-50A	0.50	1.49	1.79	2.12	2.49	2.94	3.46	4.05	4.67	5.19
M530-50A	0.50	1.64	1.96	2.33	2.74	3.23	3.79	4.42	5.06	5.59
M600-50A	0.50	2.00	2.39	2.82	3.31	3.86	4.53	5.30	6.11	6.80
M700-50A	0.50	2.29	2.72	3.21	3.76	4.39	5.14	6.00	6.90	7.66
M800-50A	0.50	2.71	3.22	3.79	4.45	5.19	6.08	7.10	8.16	9.07
M940-50A	0.50	3.09	3.68	4.33	5.07	5.92	6.94	8.10	9.31	10.34
M330-65A	0.65	1.12	1.35	1.60	1.89	2.23	2.67	3.15	3.61	4.00
M350-65A	0.65	1.17	1.41	1.67	1.97	2.31	2.75	3.23	3.69	4.07
M400-65A	0.65	1.30	1.57	1.87	2.20	2.58	3.07	3.63	4.21	4.70
M470-65A	0.65	1.48	1.79	2.12	2.49	2.92	3.45	4.06	4.69	5.22
M530-65A	0.65	1.57	1.90	2.26	2.67	3.14	3.71	4.35	5.00	5.56
M600-65A	0.65	1.82	2.19	2.61	3.08	3.62	4.25	4.95	5.68	6.30
M700-65A	0.65	2.30	2.76	3.28	3.87	4.55	5.33	6.20	7.14	8.00
M800-65A	0.65	2.58	3.09	3.67	4.32	5.08	5.94	6.90	7.93	8.87
M1000-65A	0.65	3.31	4.01	4.76	5.59	6.56	7.65	8.86	10.16	11.22
M700-100A	1.00	2.38	2.83	3.33	3.91	4.56	5.34	6.24	7.17	7.96
M800-100A	1.00	2.68	3.28	3.94	4.66	5.43	6.26	7.20	8.20	9.17

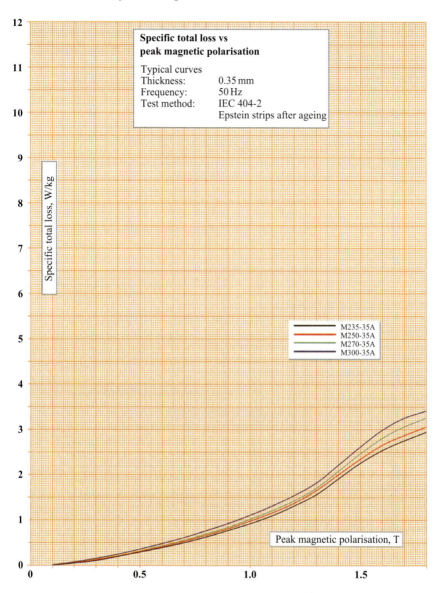

Figure 15.1

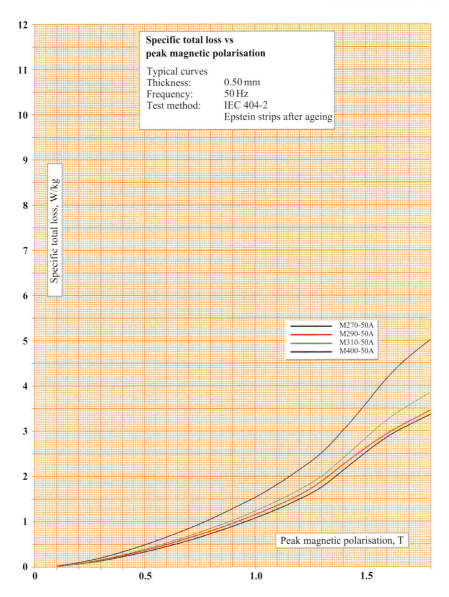

Figure 15.2

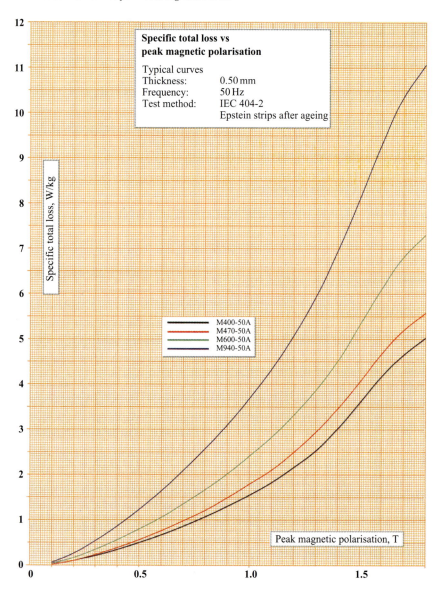

Figure 15.3

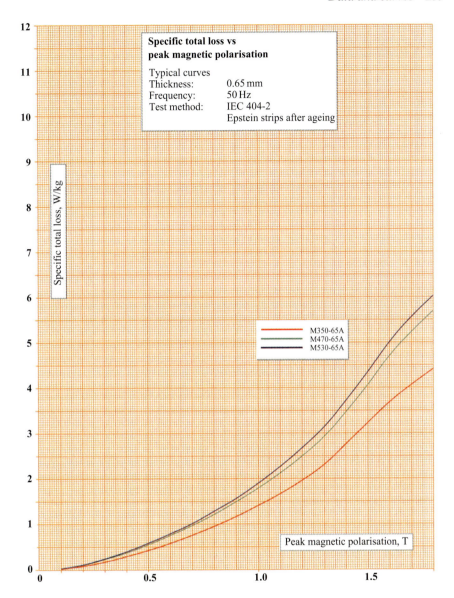

Figure 15.4

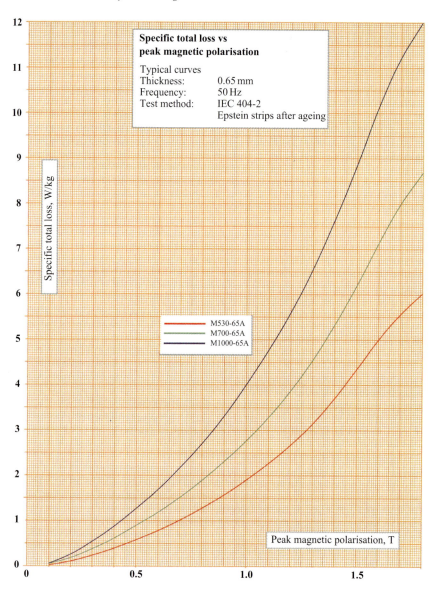

Figure 15.5

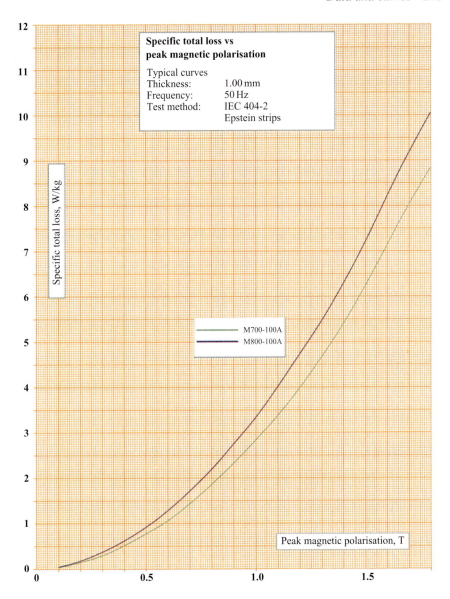

Figure 15.6

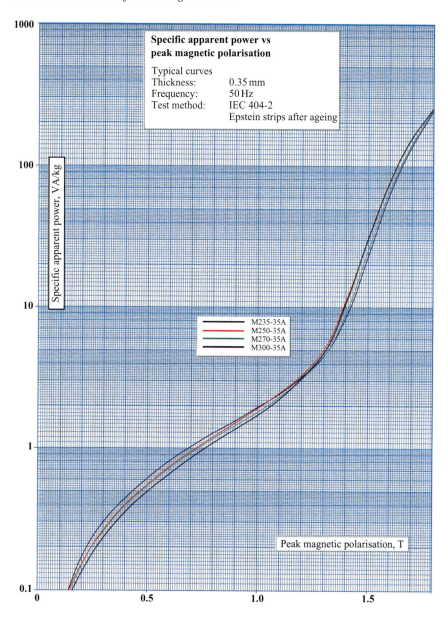

Figure 15.7

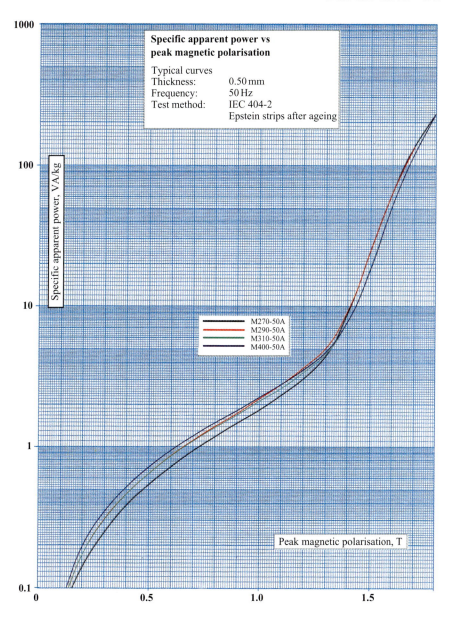

Figure 15.8

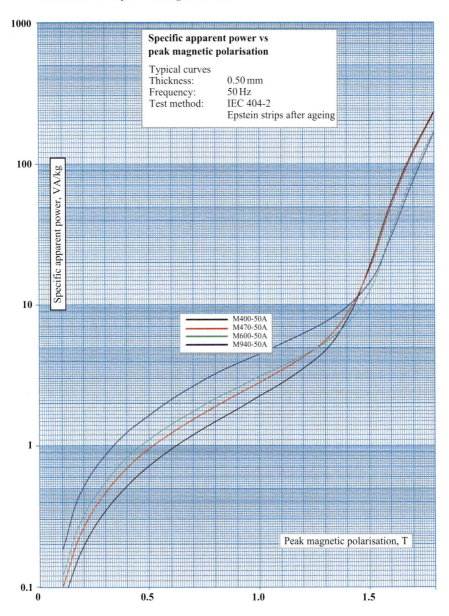

Figure 15.9

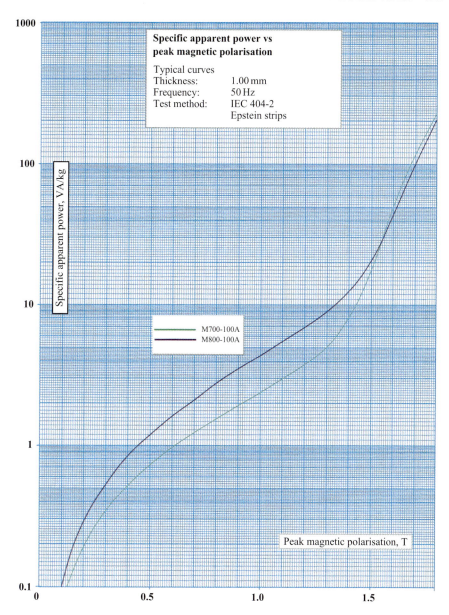

Figure 15.10

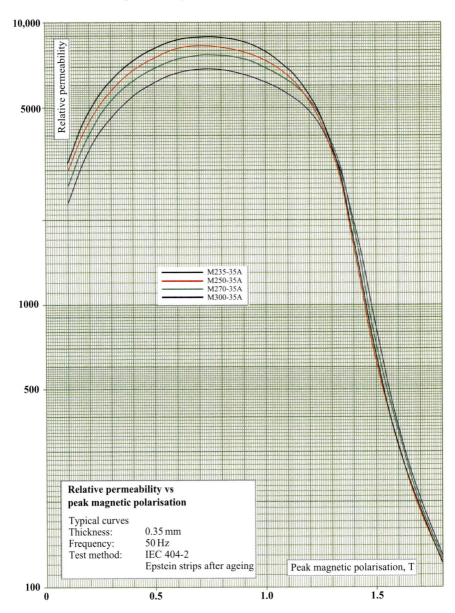

Figure 15.11

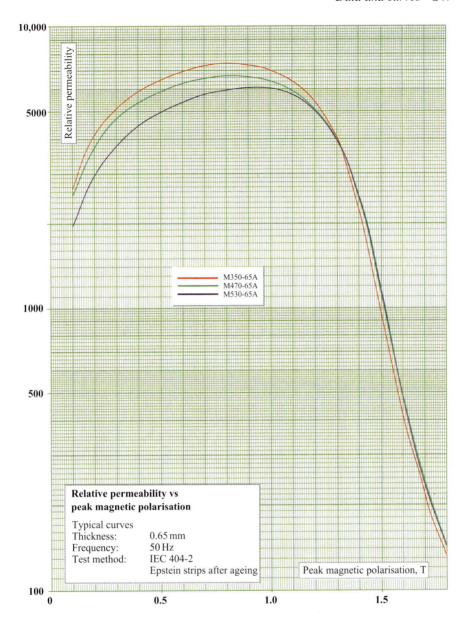

Figure 15.12

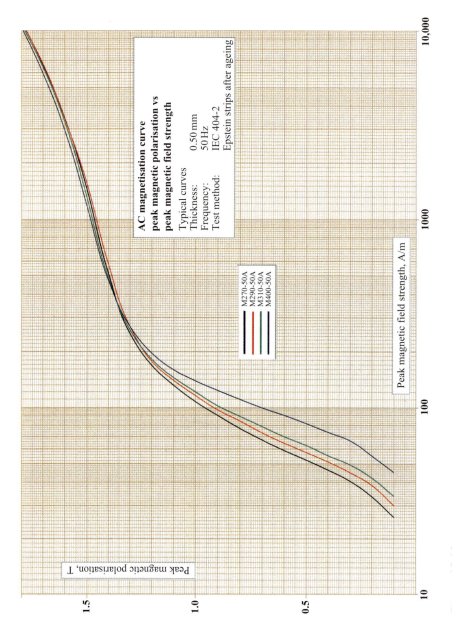

Figure 15.13

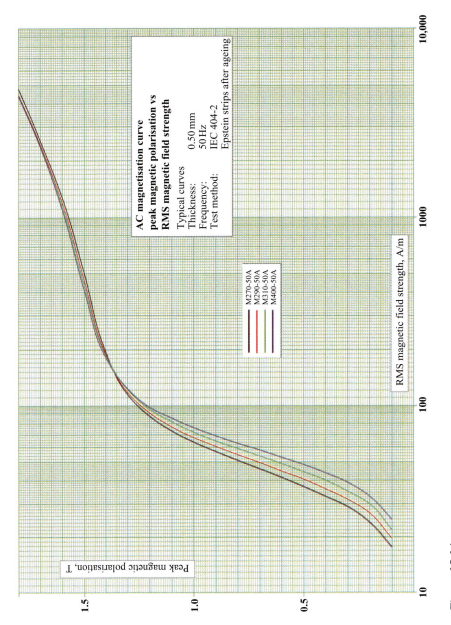

Figure 15.14

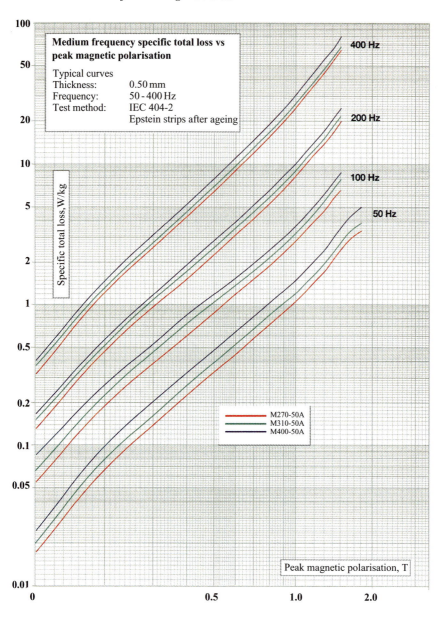

Figure 15.15

15.2   Non-oriented not finally annealed steels

Specific total loss for semi processed steels

Epstein tests

Sample sheets are taken from the beginning and end of each coil of electrical steel for magnetic tests. Epstein strips 305 mm in length and 30 mm in width are cut from each sample sheet to provide a pack of strips for measurement of the magnetic properties of that coil. Half of the pack consists of test strips taken along the rolling direction of the parent strip and half transverse to it. The Epstein pack is then given a heat treatment in a decarburising atmosphere according to that specified in the appropriate standard i.e. BS EN 10126 for the non-alloyed material and BS EN 10165 for the alloyed material. This ensures that the test specimens are in the reference condition for measurement of the magnetic properties according to the 25 cm Epstein frame method described in IEC 404-2 and BS 6404 Part 2. The following data are typical for test specimens in the reference condition.

Table 15.12   Losil alloyed electrical steels

Grade		Thickness	Conventional density	Specific total loss at $\hat{J} = 1.5$ T, 50 Hz		$\hat{J}$ at $\hat{H} = 5000$ A/m, 50 Hz		Peak relative permeability
IEC 404 8.2	BS EN 10165			Guaranteed maximum	Typical	Guaranteed minimum	Typical	Typical
		mm	kg/dm³	W/kg	W/kg	T	T	at 1.5 T
340-50-E5	M340-50E	0.50	7.65	3.40	3.10	1.62	1.64	1000
390-50-E5	M390-50E	0.50	7.70	3.90	3.20	1.64	1.66	1200
450-50-E5	M450-50E	0.50	7.75	4.50	3.70	1.65	1.69	1700
560-50-E5	M560-50E	0.50	7.80	5.60	4.15	1.66	1.70	2000
390-65-E5	M390-65E	0.65	7.65	3.90	3.60	1.62	1.64	900
450-65-E5	M450-65E	0.65	7.70	4.50	3.90	1.64	1.66	1200
520-65-E5	M520-65E	0.65	7.75	5.20	4.05	1.65	1.69	1650
630-65-E5	M630-65E	0.65	7.80	6.30	5.10	1.66	1.70	2300

Table 15.13   Newcor non-alloyed electrical steels

Grade		Thickness	Conventional density	Specific total loss at $\hat{J} = 1.5$ T, 50 Hz		$\hat{J}$ at $\hat{H} = 5000$ A/m, 50 Hz		Peak relative permeability
IEC 404 8.3	BS EN 10126			Guaranteed maximum	Typical	Guaranteed minimum	Typical	Typical
		mm	kg/dm³	W/kg	W/kg	T	T	at 1.5 T
660-50-D5	M660-50D	0.50	7.85	6.60	4.55	1.70	1.74	3000
890-50-D5	M890-50D	0.50	7.85	8.90	5.50	1.68	1.75	3000
800-65-D5	M800-65D	0.65	7.85	8.00	6.00	1.70	1.74	3000
1000-65-D5	M1000-65D	0.65	7.85	10.00	7.10	1.68	1.76	3000

Table 15.14   Polycor non-alloyed electrical steels

Grade		Thickness	Conventional density	Specific total loss at $\hat{J} = 1.5$ T, 50 Hz		$\hat{J}$ at $\hat{H} = 5000$ A/m, 50 Hz		Peak relative permeability
IEC 404 8.3	BS EN 10126			Guaranteed maximum	Typical	Guaranteed minimum	Typical	Typical
		mm	kg/dm³	W/kg	W/kg	T	T	at 1.5 T
420-50-D5*	M420-50D	0.50	7.85	4.20	3.90	1.70	1.74	3000
570-65-D5*	M570-65D	0.65	7.85	5.70	4.90	1.70	1.74	3000

*Please note that the designated grades of Polycor are not standardised grades in the IEC, BS and EN system of reference grades.

Table 15.15   Typical physical and mechanical properties

Grade	Conventional density	Resistivity	0.2% Proof stress	Ultimate tensile strength	Elongation (80 mm gauge	Hardness	Stacking factor	Bend test
	kg/dm^3	$\mu\Omega$cm	N/mm^2	N/mm^2	length) %	VPN	%	
Losil								
M340-50E	7.65	42	480	560	11	210	97	>10
M390-50E	7.70	37	450	530	15	200	97	>10
M450-50E	7.75	30	430	510	16	190	97	>10
M560-50E	7.80	22	420	500	16	180	97	>10
M390-65E	7.65	42	480	560	11	210	97	>10
M450-65E	7.70	37	450	530	15	200	97	>10
M520-65E	7.75	30	430	510	16	190	97	>10
M630-65E	7.80	22	400	480	20	170	97	>10
Newcor								
M660-50D	7.85	17	450	530	14	180	97	>10
M890-50D	7.85	14	440	520	15	170	97	>10
M800-65D	7.85	17	450	530	14	180	97	>10
M1000-65D	7.85	14	350	450	20	160	97	>10
Polycor								
M420-50D	7.85	22	390	420	16	140	97	>10
M570-65D	7.85	22	390	420	16	140	97	>10

Summit
Summit is a non-guaranteed grade of electrical steel produced in the semi-processed condition at thicknesses from 0.50 mm to 1.00 mm. It is available in two grades: Summit 125 and Summit 140 which have typical hardnesses (VPN) of 125 and 140 respectively.

Note: Vickers hardness tests are carried out in accordance with BS 427: Part 1:1961 (1981) using a load of 10 kgf or 5 kgf according to thickness and hardness range.
The above physical and mechanical data refer to material in the as delivered condition i.e. the semi-processed state.

Surface condition

The products described are normally supplied in the uncoated condition. The surface condition, and in particular the surface roughness of the material, can be subject to agreement. Alternatively, Polycor can be delivered with an insulation coating on both surfaces. The coating is essentially inorganic with resin additions and can withstand annealing. A reduction in the surface insulation resistance is found on annealing. The coating has good welding characteristics with low gassing over the region of the weld.

Specific total loss, W/kg at 50 Hz

Table 15.16   *Losil alloyed electrical steels*

Grade		Specific total loss at frequency of 50 Hz (W/kg) $\hat{J}(T) =$								
IEC 404 404 8.2	BS EN 10165	0.3	0.5	0.7	0.9	1.1	1.3	1.5	1.7	1.8
340-50-E5	M340-50E	0.150	0.354	0.650	1.020	1.47	2.08	3.08	4.23	4.67
390-50-E5	M390-50E	0.161	0.386	0.693	1.077	1.54	2.17	3.20	4.32	4.80
450-50-E5	M450-50E	0.192	0.455	0.814	1.260	1.83	2.59	3.70	5.07	5.69
560-50-E5	M560-50E	0.213	0.512	0.910	1.430	2.10	2.94	4.14	5.60	6.20
390-65-E5	M390-65E	0.197	0.440	0.770	1.175	1.69	2.46	3.57	4.75	5.30
450-65-E5	M450-65E	0.233	0.494	0.865	1.320	1.91	2.70	3.93	5.34	5.86
520-65-E5	M520-65E	0.220	0.500	0.915	1.460	2.12	2.98	4.04	5.23	5.75
630-65-E5	M630-65E	0.270	0.610	1.090	1.70	2.50	3.55	5.10	6.75	7.40

Table 15.17   *Newcor non-alloyed electrical steels*

Grade		Specific total loss at frequency of 50 Hz (W/kg) $\hat{J}(T) =$								
IEC 404 404 8.3	BS EN 10126	0.3	0.5	0.7	0.9	1.1	1.3	1.5	1.7	1.8
660-50-D5	M660-50D	0.268	0.610	1.075	1.662	2.380	3.29	4.57	6.20	6.75
890-50-D5	M890-50D	0.320	0.732	1.283	1.980	2.881	4.02	5.60	7.44	8.08
800-65-D5	M800-65D	0.300	0.714	1.300	2.054	3.020	4.30	6.06	8.08	8.85
1000-65-D5	M1000-65D	0.357	0.847	1.530	2.474	3.670	5.17	7.10	9.34	10.35

Table 15.18   *Polycor non-alloyed electrical steels*

Grade		Specific total loss at frequency of 50 Hz (W/kg) $\hat{J}(T) =$								
IEC 404 404 8.3	BS EN 10126	0.3	0.5	0.7	0.9	1.1	1.3	1.5	1.7	1.8
420-50-D5	M420-50D	0.218	0.506	0.898	1.350	2.02	2.77	3.85	5.28	5.75
570-65-D5	M570-65D	0.237	0.585	1.081	1.730	2.59	3.62	4.95	6.65	7.30

Specific total loss, W/kg at 60 Hz

Table 15.19   Losil alloyed electrical steels

Grade		Thickness	Specific total loss at frequency of 60 Hz (W/kg)	
IEC 404 8.2	BS EN 10165	mm	Guaranteed maximum	Typical
340-50-E5	M340-50E	0.50	4.32	3.95
390-50-E5	M390-50E	0.50	4.97	4.10
450-50-E5	M450-50E	0.50	5.67	4.70
560-50-E5	M560-50E	0.50	7.03	5.30
390-65-E5	M390-65E	0.65	5.07	4.60
450-65-E5	M450-65E	0.65	5.86	5.00
520-65-E5	M520-65E	0.65	6.72	5.25
630-65-E5	M630-65E	0.65	8.00	6.50

Table 15.20   Newcor non-alloyed electrical steels

Grade		Thickness	Specific total loss at frequency of 60 Hz (W/kg)	
IEC 404 8.3	BS EN 10126	mm	Guaranteed maximum	Typical
660-50-D5	M660-50D	0.50	8.38	5.80
890-50-D5	M890-50D	0.50	11.30	7.00
800-65-D5	M800-65D	0.65	10.16	7.65
1000-65-D5	M1000-65D	0.65	12.70	9.05

Table 15.21   Polycor non-alloyed electrical steels

Grade		Thickness	Specific total loss at frequency of 60 Hz (W/kg)	
IEC 404 8.3	BS EN 10126	mm	Guaranteed maximum	Typical
420-50-D5*	M420-50D	0.50	5.33	4.95
570-65-D5*	M570-65D	0.65	7.24	6.25

*Please note that the designated grades of Polycor are not standardised grades in the IEC, BS and EN system of reference grades.

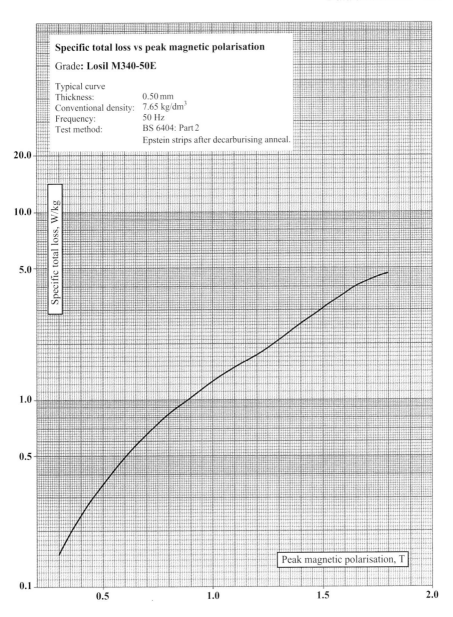

Specific total loss vs peak magnetic polarisation

Grade: **Losil M340-50E**

Typical curve
Thickness: 0.50 mm
Conventional density: 7.65 kg/dm³
Frequency: 50 Hz
Test method: BS 6404: Part 2
                           Epstein strips after decarburising anneal.

Figure 15.16

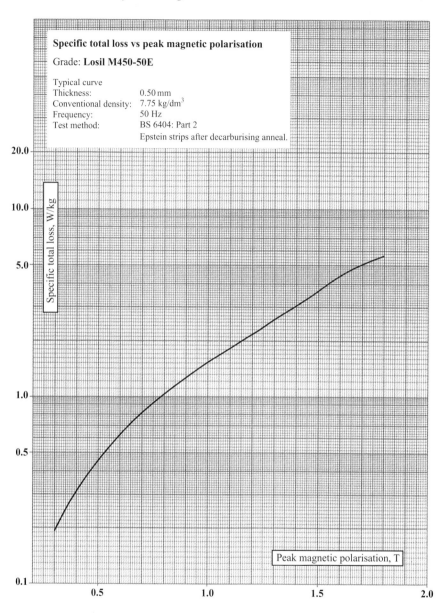

Specific total loss vs peak magnetic polarisation

Grade: **Losil M450-50E**

Typical curve
Thickness: 0.50 mm
Conventional density: 7.75 kg/dm^3
Frequency: 50 Hz
Test method: BS 6404: Part 2
                        Epstein strips after decarburising anneal.

Figure 15.17

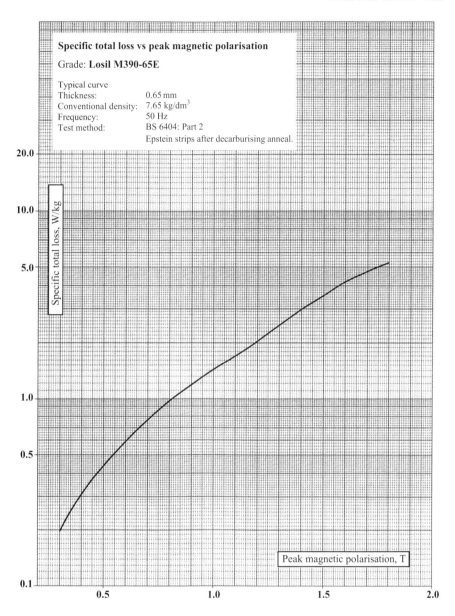

Figure 15.18

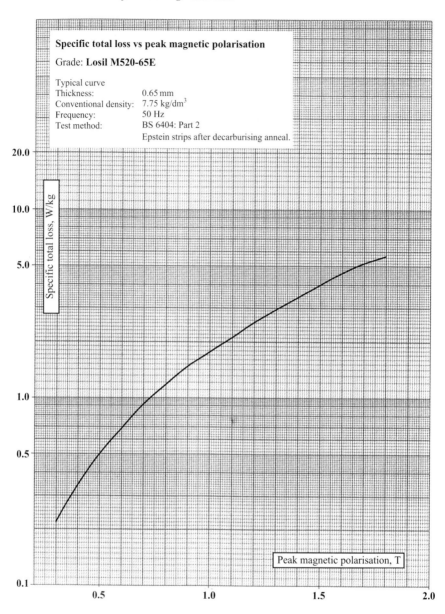

Figure 15.19

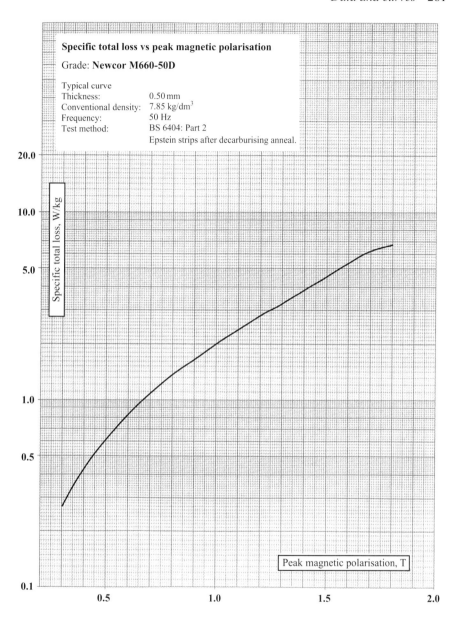

Figure 15.20

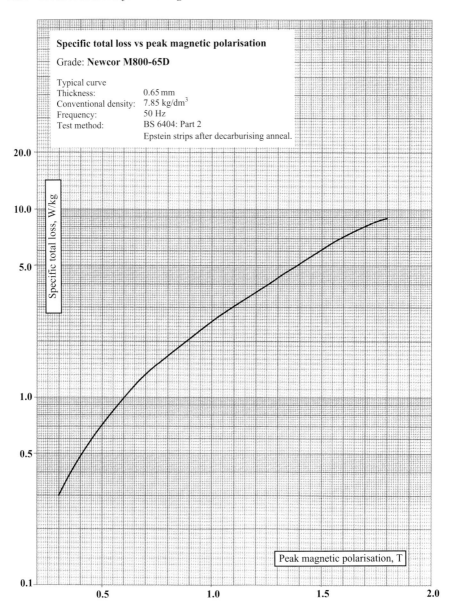

Specific total loss vs peak magnetic polarisation

Grade: **Newcor M800-65D**

Typical curve
Thickness: 0.65 mm
Conventional density: 7.85 kg/dm³
Frequency: 50 Hz
Test method: BS 6404: Part 2
                          Epstein strips after decarburising anneal.

Specific total loss, W/kg

Peak magnetic polarisation, T

Figure 15.21

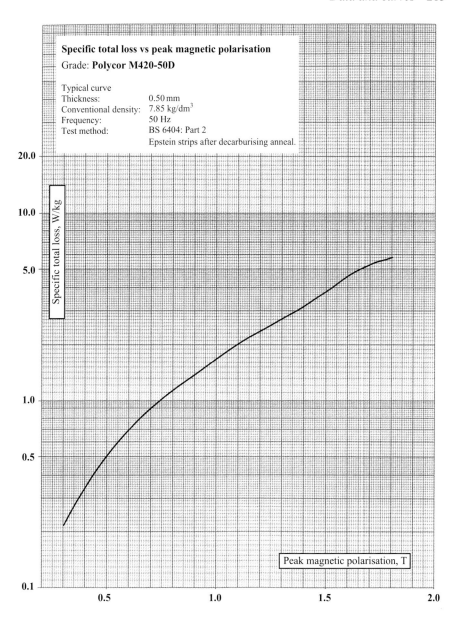

Specific total loss vs peak magnetic polarisation

Grade: **Polycor M420-50D**

Typical curve
Thickness: 0.50 mm
Conventional density: 7.85 kg/dm³
Frequency: 50 Hz
Test method: BS 6404: Part 2
Epstein strips after decarburising anneal.

Specific total loss, W/kg

Peak magnetic polarisation, T

Figure 15.22

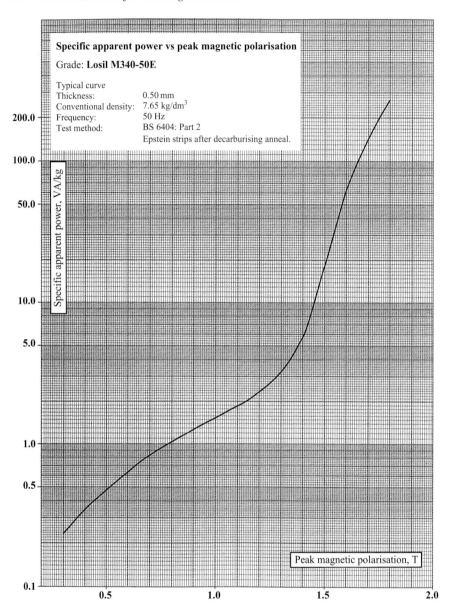

Figure 15.23

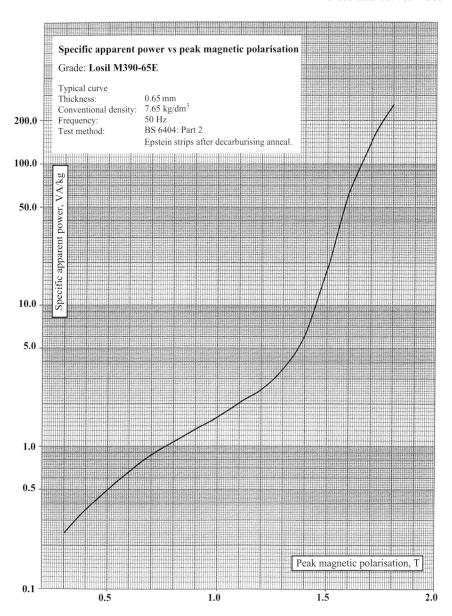

Figure 15.24

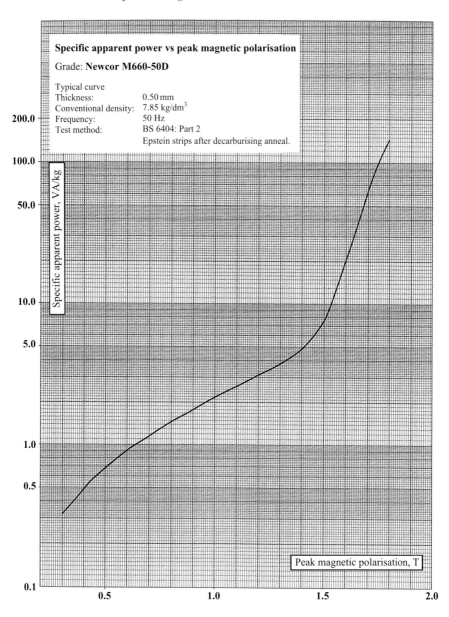

Figure 15.25

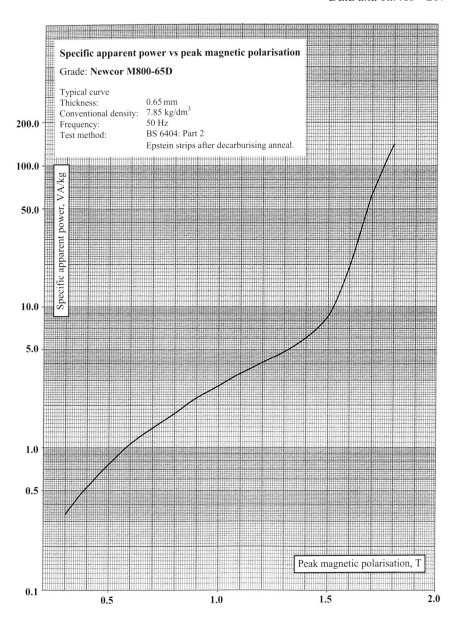

Specific apparent power vs peak magnetic polarisation

Grade: **Newcor M800-65D**

Typical curve
Thickness: 0.65 mm
Conventional density: 7.85 kg/dm³
Frequency: 50 Hz
Test method: BS 6404: Part 2
                         Epstein strips after decarburising anneal.

Specific apparent power, V A/kg

Peak magnetic polarisation, T

Figure 15.26

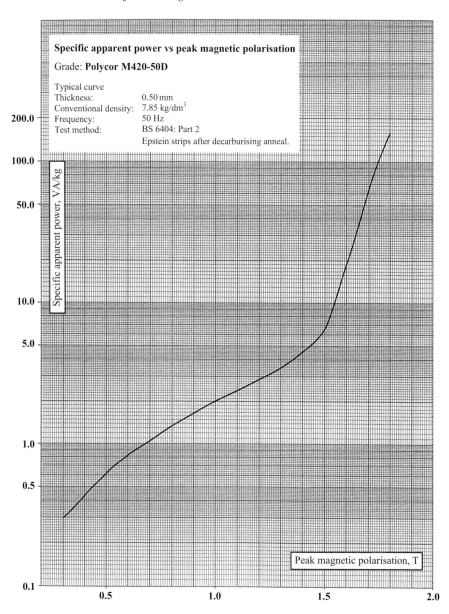

Specific apparent power vs peak magnetic polarisation

Grade: **Polycor M420-50D**

Typical curve
Thickness: 0.50 mm
Conventional density: 7.85 kg/dm^3
Frequency: 50 Hz
Test method: BS 6404: Part 2
Epstein strips after decarburising anneal.

Specific apparent power, V A/kg

Peak magnetic polarisation, T

Figure 15.27

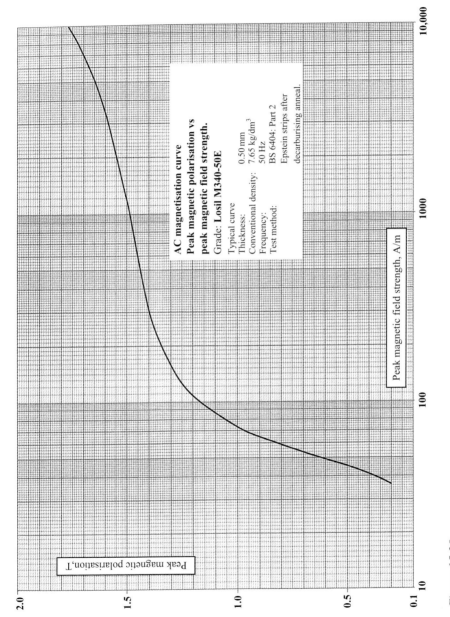

AC magnetisation curve
Peak magnetic polarisation vs
peak magnetic field strength.
Grade: **Losil M340-50E**

Typical curve	
Thickness:	0.50 mm
Conventional density:	7.65 kg/dm^3
Frequency:	50 Hz
Test method:	BS 6404: Part 2
	Epstein strips after
	decarburising anneal.

Peak magnetic field strength, A/m

Peak magnetic polarisation, T

Figure 15.28

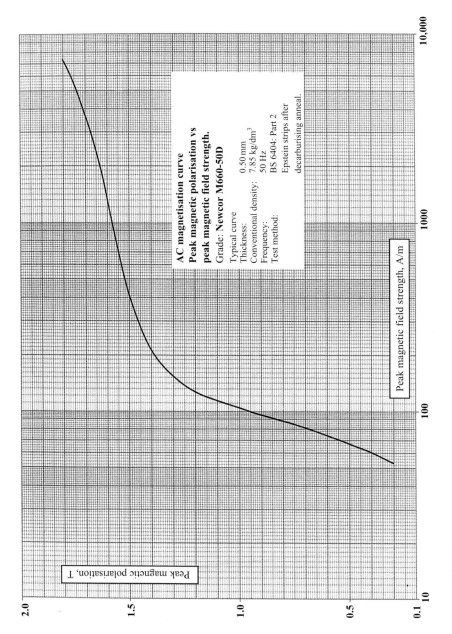

Figure 15.29

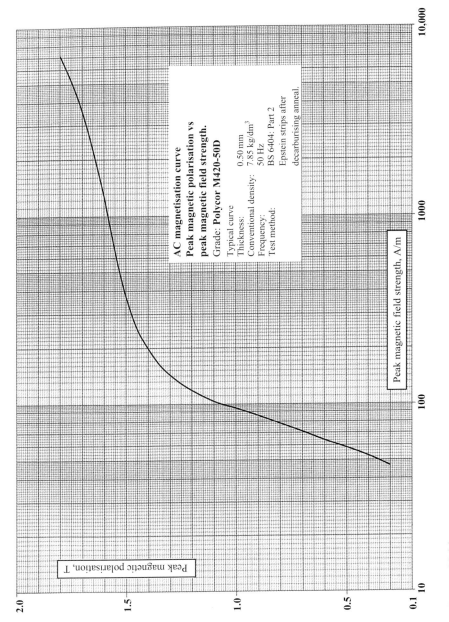

AC magnetisation curve
**Peak magnetic polarisation vs
peak magnetic field strength.**
Grade: **Polycor M420–50D**

Typical curve	
Thickness:	0.50 mm
Conventional density:	7.85 kg/dm³
Frequency:	50 Hz
Test method:	BS 6404: Part 2
	Epstein strips after
	decarburising anneal.

Peak magnetic field strength, A/m

Peak magnetic polarisation, T

Figure 15.30

15.3   Grain-oriented steels

Specific total loss

Epstein tests

Grain oriented electrical steels are graded according to the specifications of the International Electrotechnical Commission, Publication IEC 404-8-7, the British Standard BS 6404: Section 8.7 and the European Standard EN 10107. As set out in these standards the measurement of specific total loss is performed according to the 25 cm Epstein frame method described in IEC 404-2 and BS 6404: Part 2.

Equivalent 'M' grades which were based on British Standard 601 have now been withdrawn and have been replaced by the IEC/EN designations. The nearest equivalent grades are given for information purposes only.

Table 15.22

| Grade | | Thickness | Specific total loss at frequency of 50 Hz (W/kg) | | | | $\hat{B}$ at $\hat{H}=$ 800 A/m, 50 Hz |
| IEC 404 BS 6404 | EN 10107 | | Guaranteed max. | | Typical | | Typical |
		mm	$\hat{B} = 1.5\,\text{T}$	$1.7\,\text{T}$	$1.5\,\text{T}$	$1.7\,\text{T}$	T
Unisil-H							
103-27-P5	M103-27P	0.27	–		1.03	0.74	1.00	1.93
105-30-P5	M105-30P	0.30	–		1.05	0.77	1.03	1.93
111-30-P5	M111-30P	0.30	–		1.11	0.80	1.08	1.93
117-30-P5	M117-30P	0.30	–		1.17	0.84	1.14	1.92
Unisil							
120-23-S5	M120-23S	0.23	0.77	1.20	0.73	1.13	1.83
080-23-N5	M080-23N	0.23	0.80	1.27	0.76	1.15	1.83
130-27-S5	M130-27S	0.27	0.85	1.30	0.79	1.16	1.83
089-27-N5	M089-27N	0.27	0.89	1.40	0.83	1.21	1.83
140-30-S5	M140-30S	0.30	0.92	1.40	0.85	1.22	1.83
097-30-N5	M097-30N	0.30	0.97	1.50	0.91	1.31	1.83
155-35-S5	M150-35S	0.35	1.05	1.50	0.98	1.38	1.83
111-35-N5	M111-35N	0.35	1.11	1.65	1.02	1.43	1.83
175-50-N5*	M175-50N	0.50	1.75	–	1.35	1.92	1.82

Orb Electrical Steels is currently developing 0.23 mm high permeability grain oriented silicon steel and domain controlled high permeability grain oriented silicon steel.

*Please note that this grade does not appear in the IEC, BS or EN standards.

Table 15.23

Old designation	New designation EN 10107
27M0H	M103-27P
30M0H	M105-30P
30M1H	M111-30P
30M2H	M117-30P

Note: Data for the frequency of 60 Hz are given on page 277.

Table 15.24

Old designation	New designation EN 10107
23M3	M080-23N
27M3	M130-27S
27M4	M089-27N
30M5	M097-30N
35M6	M111-35N
50M7	M175-50N

Single sheet tests

The specific total loss of grain oriented electrical steels may increase if the steel sheet retains a degree of internal stress after final processing. These stresses may be removed by subjecting the cut transformer laminations to a stress relief anneal. However, the modem production processes and controls for Unisil and Unisil-H realise a product with very low levels of internal stress. The material has loss characteristics which are so close to its best potential value that annealing of wide strip by the customer is becoming less common.

Specific total loss values based on annealed Epsteins may be less relevant to the user than losses measured on single sheets which have been tested in the non-stress relief annealed condition. Therefore, Orb Electrical Steels also samples and tests all coils of Unisil and Unisil-H in the single sheet non-annealed condition.

The method of measurement of the power loss in single sheet test specimens has been the subject of considerable debate in the technical committees of the International Electrotechnical Commission (IEC). The publication IEC 404-3: 1922 (and the technically related BS 6404: Part 3 1992) represents the latest consensus of world-wide opinion on the subject of single sheet testing. The method utilises test specimens of size 500 mm × 500 mm. It is recognised that, in the past, it has frequently been the case that the calibration of single sheet test systems was by direct correlation with the Epstein frame. In the case of the IEC test system, however, the measurement of power loss by the single sheet method is entirely independent of the Epstein frame. Accordingly, the results from the single sheet tester may differ from those of the Epstein frame by about 5%. Typical power loss values obtained by measurements carried out according to the method of IEC 404-3: 1992 are given in the following table:

Table 15.25   Typical single sheet test data

Grade		Thickness	Specific total loss at frequency of 50 Hz (W/kg)	
IEC 404 BS 6404	EN 10107		Typical	
		mm	$\hat{B} = 1.5\,\text{T}$	1.7 T
Unisil-H				
103-27-P5	M103-27P	0.27	0.77	1.06
105-30-P5	M105-30P	0.30	0.80	1.09
111-30-P5	M111-30P	0.30	0.84	1.15
117-30-P5	M117-30P	0.30	0.88	1.21
Unisil				
120-23-S5	M120-23S	0.23	0.76	1.20
080-23-N5	M080-23N	0.23	0.79	1.22
130-27-S5	M130-27S	0.27	0.82	1.23
089-27-N5	M089-27N	0.27	0.87	1.28
140-30-S5	M140-30S	0.30	0.89	1.30
097-30-N5	M097-30N	0.30	0.95	1.39
155-35-S5	M150-35S	0.35	1.02	1.46
111-35-N5	M111-35N	0.35	1.06	1.52
175-50-N5*	M175-50N	0.50	1.41	2.04

*Please note that this grade does not appear in the IEC, BS or EN standards.

Table 15.26 Typical physical properties

	Unisil-H	Unisil
Density, kg/dm^3	7.65	7.65
Silicon content, %	2.90	3.10
Resistivity, microhm cm	45	48
0.1% Proof stress, N/mm^2 (kg/mm^2)		
0° to rolling direction	300(30.6)	300(30.6)
90° to rolling direction	315(32.1)	308(31.4)
Ultimate tensile strength, N/mm^2 (kg/mm^2)		
0° to rolling direction	325(33.1)	320(32.6)
90° to rolling direction	385(39.2)	375(38.2)
% Elongation on 80 mm gauge length		
0° to rolling direction	11	6
90° to rolling direction	33	36
Hardness, HV 2.5 kg	175	175
Bend test	>6	>6
Stacking factor, %		
0.23 mm	N.A.	96.0
0.27 mm	96.0	96.0
0.30 mm	96.5	96.5
0.35 mm	N.A.	97.0
0.50 mm	N.A.	97.0

Table 15.27   Typical specific total loss data, W/kg at 50 Hz

Grade		Specific total loss at frequency of 50 Hz (W/kg) $\hat{B}$ (T)=							
IEC 404 BS 6404	EN 10107	0.5	0.7	0.9	1.1	1.3	1.5	1.7	1.9
Unisil-H									
103-27-P5	M103-27P	0.112	0.192	0.294	0.416	0.565	0.740	1.00	1.63
105-30-P5	M105-30P	0.114	0.197	0.300	0.425	0.585	0.768	1.03	1.70
111-30-P5	M111-30P	0.118	0.210	0.321	0.452	0.615	0.800	1.08	1.75
117-30-P5	M117-30P	0.126	0.220	0.338	0.475	0.640	0.842	1.14	1.85
Unisil									
120-23-S5	M120-23S		0.161	0.260	0.382	0.527	0.730	1.13	1.89
080-23-N5	M080-23N		0.168	0.270	0.392	0.549	0.761	1.15	1.89
130-27-S5	M130-27S		0.175	0.284	0.420	0.581	0.790	1.16	1.90
089-27-N5	M089-27N	0.100	0.178	0.290	0.427	0.596	0.830	1.21	1.90
140-30-S5	M140-30S	0.101	0.191	0.311	0.453	0.624	0.850	1.22	1.90
097-30-N5	M097-30N	0.117	0.217	0.343	0.495	0.676	0.910	1.31	1.94
155-35-S5	M150-35S	0.130	0.239	0.374	0.530	0.725	0.980	1.38	2.08
111-35-N5	M111-35N	0.142	0.258	0.395	0.561	0.765	1.02	1.43	2.10
175-50-N5	M175-50N	0.164	0.302	0.485	0.710	0.985	1.35	1.92	2.70

Table 15.28 Specific total loss data, W/kg at 60 Hz

| Grade | | Thickness | Specific total loss at frequency of 60 Hz (W/kg) | | | |
| IEC 404 BS 6404 | EN 10107 | | Guaranteed max. | | Typical | |
		mm	$\hat{B} = 1.5\,T$	1.7 T	1.5 T	1.7 T
Unisil-H						
103-27-P5	M103-27P	0.27	–	1.35	0.97	1.30
105-30-P5	M105-30P	0.30	–	1.38	1.00	1.35
111-30-P5	M111-30P	0.30	–	1.46	1.05	1.41
117-30-P5	M117-30P	0.30	–	1.54	1.10	1.49
Unisil						
120-23-S5	M120-23S	0.23	1.01	1.57	0.95	1.48
080-23-N5	M080-23N	0.23	1.06	1.65	0.99	1.50
130-27-S5	M130-27S	0.27	1.12	1.68	1.03	1.51
089-27-N5	M089-27N	0.27	1.17	1.85	1.08	1.57
140-30-S5	M140-30S	0.30	1.21	1.83	1.10	1.60
097-30-N5	M097-30N	0.30	1.28	1.98	1.18	1.70
155-35-S5	M150-35S	0.35	1.38	1.98	1.30	1.81
111-35-N5	M111-35N	0.35	1.46	2.18	1.33	1.86
175-50-N5	M175-50N	0.50	2.27	–	1.89	2.65

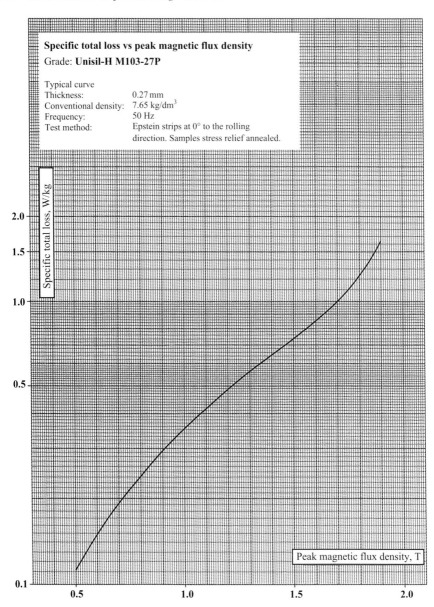

Specific total loss vs peak magnetic flux density
Grade: **Unisil-H M103-27P**

Typical curve
Thickness: 0.27 mm
Conventional density: 7.65 kg/dm³
Frequency: 50 Hz
Test method: Epstein strips at 0° to the rolling
                       direction. Samples stress relief annealed.

Specific total loss, W/kg

Peak magnetic flux density, T

Figure 15.31

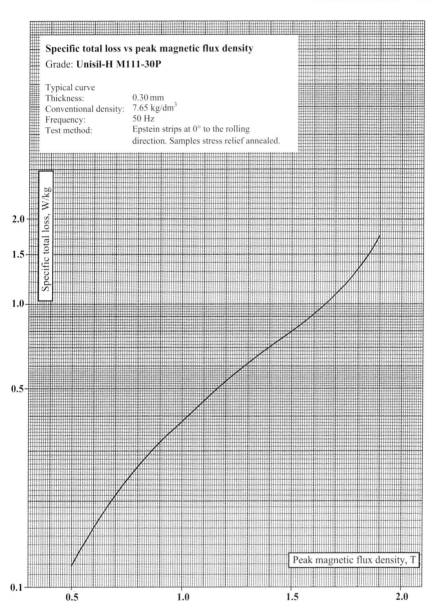

Specific total loss vs peak magnetic flux density

Grade: **Unisil-H M111-30P**

Typical curve
Thickness: 0.30 mm
Conventional density: 7.65 kg/dm³
Frequency: 50 Hz
Test method: Epstein strips at 0° to the rolling
                        direction. Samples stress relief annealed.

Figure 15.32

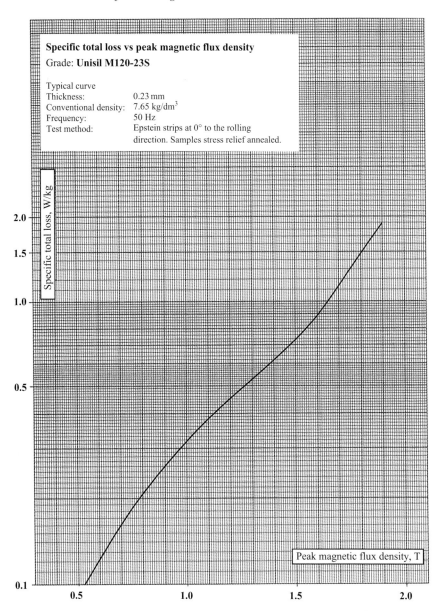

Figure 15.33

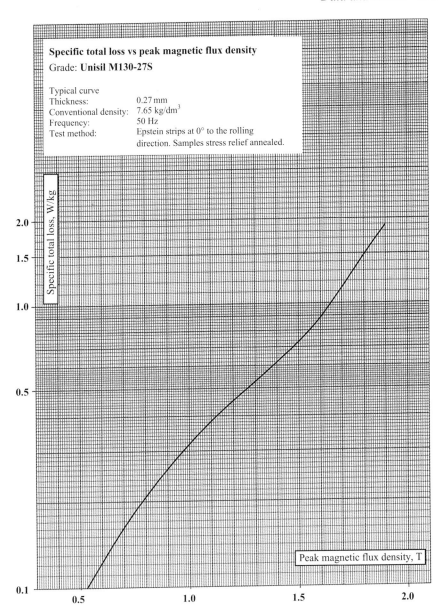

Specific total loss vs peak magnetic flux density

Grade: **Unisil M130-27S**

Typical curve
Thickness: 0.27 mm
Conventional density: 7.65 kg/dm^3
Frequency: 50 Hz
Test method: Epstein strips at 0° to the rolling
                        direction. Samples stress relief annealed.

Specific total loss, W/kg

Peak magnetic flux density, T

Figure 15.34

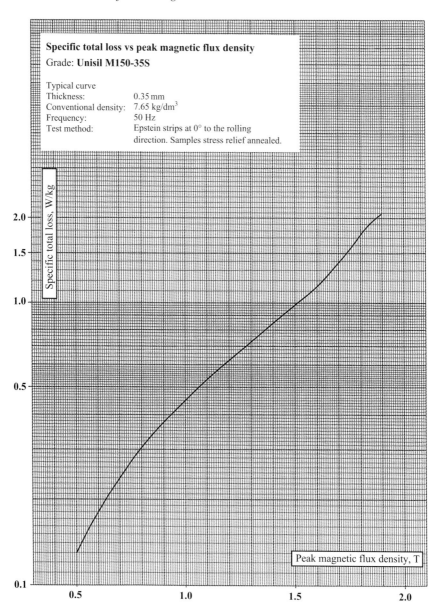

Specific total loss vs peak magnetic flux density
Grade: **Unisil M150-35S**

Typical curve
Thickness: 0.35 mm
Conventional density: 7.65 kg/dm³
Frequency: 50 Hz
Test method: Epstein strips at 0° to the rolling
                          direction. Samples stress relief annealed.

Figure 15.35

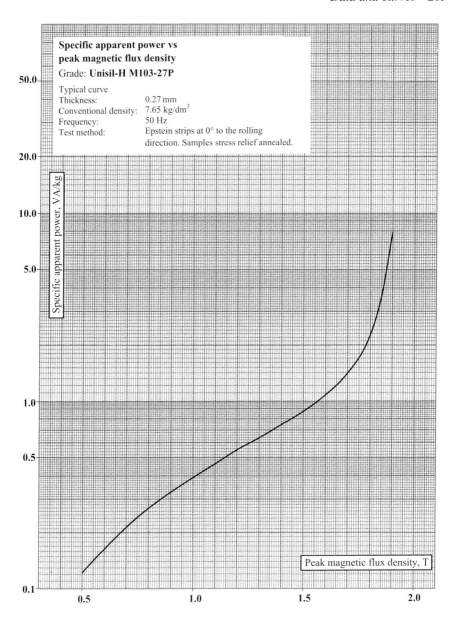

Figure 15.36

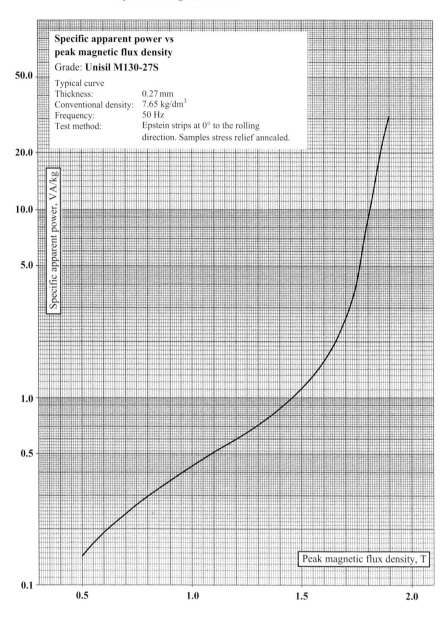

Figure 15.37

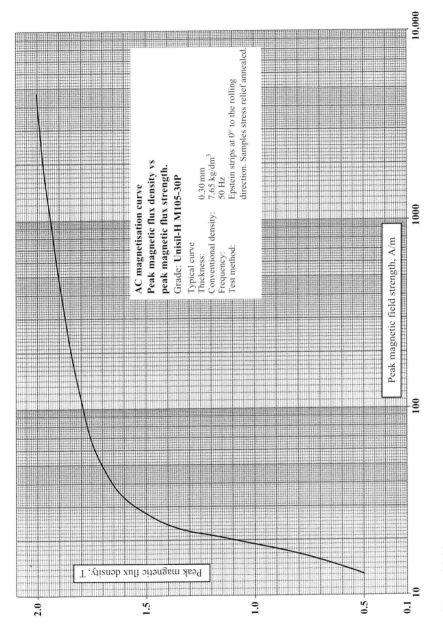

AC magnetisation curve
Peak magnetic flux density vs peak magnetic flux strength.
Grade: **Unisil-H M105-30P**

Typical curve	
Thickness:	0.30 mm
Conventional density:	7.65 kg/dm³
Frequency:	50 Hz
Test method:	Epstein strips at 0° to the rolling direction. Samples stress relief annealed.

Peak magnetic field strength, A/m

Peak magnetic flux density, T

Figure 15.38

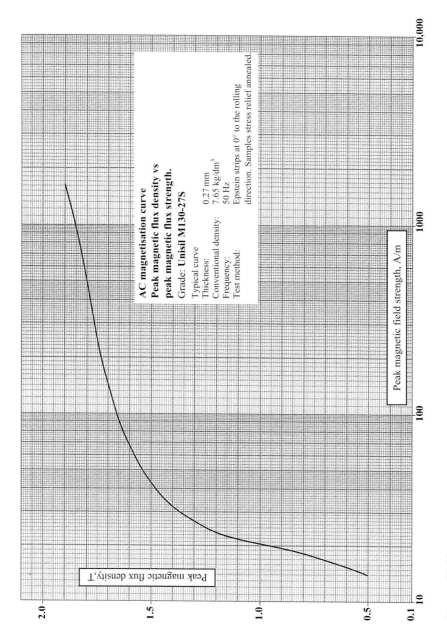

AC magnetisation curve
**Peak magnetic flux density vs
peak magnetic flux strength.**
Grade: **Unisil M130-27S**

Typical curve
Thickness: 0.27 mm
Conventional density: 7.65 kg/dm^3
Frequency: 50 Hz
Test method: Epstein strips at 0° to the rolling
                           direction. Samples stress relief annealed

Peak magnetic field strength, A/m

Peak magnetic flux density, T

Figure 15.39

Appendix
Glossary

ageing The progressive change in properties, magnetic or mechanical, which takes place over time. Ageing is accelerated at higher temperatures, e.g. 150 °C. Ageing is associated with the progressive precipitation of species in solid solution, e.g. carbides, nitrides.

ampère (amp) Unit of electric current, defined in terms of the force between adjacent current-carrying conductors or as being that current which releases 0.001118 g of silver in one second in a silver voltameter.

anisotropic Property of material or field such that it has dissimilar characteristics in different directions. Example: grain-oriented electrical steel has superior magnetic properties in its rolling direction.

anti-ageing Some additives to steel, e.g. titanium, react with carbon in solution and restrain carbide precipitates. Anti-ageing agents may have adverse side-effects.

autoclave An oven or small furnace able to be sealed against the atmosphere allowing items to be heated in an oxygen-free environment, e.g. motors to be freed of epoxy-glued-in windings for repair.

average The mean value of a property, particularly the time average of dB/dt signals give a route to peak induction values.

'A' weighting In acoustics 'A' weighting describes a correction applied to sound levels which normalise them over a wide frequency range to compensate for the human ear's variation of sensitivity with frequency.

back iron In a motor or alternator, stator windings are placed around 'teeth' to form the inward facing poles of the machine. The magnetic circuit between teeth is completed by an outer annulus of steel known as back iron. It can be made deep to reduce circuit magnetic reluctance and reduce losses at the expense of increased steel usage.

ballistic galvanometer When a pulse of electricity flows through an electric measuring instrument the response of the electromechanical part of the instrument may be slow compared to the length of the pulse. However the mechanical impulse is absorbed by the moveable component and translated into a measured deflection against a known restoring force. The response to impulse is labelled 'ballistic'.

B coil   This is an array of conductors enclosing a piece of magnetic material (or air) and able to generate an emf at its terminals proportional to the time derivative of the field concerned, i.e. $V_{out} \propto dB/dt$. Flux measuring systems involve B coils which are in fact dB/dt coils to be precise.

beta rays β-rays are high-speed electrons emitted during nuclear decay of certain isotopes, e.g. Sr 90. The absorption of beta-rays has been used as the basis of a thickness measuring device for steel.

bridge methods   The power losses in a circuit containing electrical steel may be represented as an equivalent resistance in an electrical circuit. Permeability appears as the increased level of inductance produced. By including a steel sample along with its winding in a bridge circuit, balances can be established to determine the resistive and reactive components involved. Chapter 10 discusses bridge methods and refers to Hague's book on this topic.

carbon black   When gas is burned in a limited oxygen supply some unreacted carbon is produced in the form of soot or carbon black. This is a valuable industrial product but is undesired and a sign of incorrect operation of processes designed to produce annealing atmospheres.

catenary furnace   This is a form of strand annealing furnace in which strip hangs in a long catenary suspended between input and output rolls. No positive tension is applied as strip passes through but the weight of the catenary produces some slight stretching and flattening of the strip. The strip has to be hot enough for the creep rate in the steel to be high enough to achieve flattening during the residence time of the strip.

cold work   This normally refers to a rolling process and indicates that permanent deformation beyond the elastic limit of steel has taken place at a temperature which is below that at which re-crystallisation may occur.

commutator   This is part of a motor or dynamo which switches between rotor coils as the machine rotor turns so that current flows in appropriate directions through the coils. In the case of dynamos it amounts to being a mechanical rectifier.

corrosion inhibitor   This is an additive put into cold rolling lubricant fluid so that steel strip is protected from early rusting during limited storage. A wide range of organic components is used. Also wrapping paper may be impregnated with related compounds to protect wrapped steel during transit.

coulomb   A unit of quantity of electricity. 1 ampère represents 1 coulomb per second passing along a conductor ($= 6.25 \times 10^{18}$ electrons/second).

cracking   The thermal decomposition of a compound, usually a gas or oil.

creep rate   When a hot metal is exposed to stress it will slowly deform (creep). The creep rate is the time required at a given temperature and stress level for significant deformation to occur in the time taken by the processes involved.

crown Rolling tends naturally to deliver strip thicker in the centre than at the edges. The ends of rolls are less liable to deformation than the mid-parts. The degree of overthickness at the centre of the strip is called the crown. It is progressive from edge to centre. Many rolling strategies exist to minimise the presence of crown.

Curie point This is the temperature at which ferromagnetic effects finally vanish after decreasing progressively as temperature is raised. At the Curie point thermal disruption overcomes the tendency for spin alignments to form in a coherent manner.

customer anneal This is an annealing treatment of whatever form applied by the end-user of electrical steel rather than the manufacturer. It is normally applied after final forming and may be decarburising, provocative of critical grain growth or merely stress relieving.

cycloconverter This is a power electronics system in which power is absorbed at mains supply frequency (50 or 60 Hz) and delivered at variable voltage and (usually) higher frequency for the operation of speed-controlled motors. Multiphase arrangements of thyristors or force-commutated semiconductors are used.

Dannatt plates These are copper plates contained within magnetising coils and placed parallel to steel strip being magnetised. Any component of applied field which tends to appear normal to the surface of these plates invokes a cancelling (opposing) field, due to Lenz law action, of the eddy current produced in the copper. They have the effect of improving the uniformity of an effective field and the consequent magnetisation of strip in a magnetiser.

delta/star In three-phase systems supply lines and machine windings may be set out in delta or star formation between points P_1, P_2, P_3. The voltage P_1–P_3 in a delta system is $\sqrt{3}$ times the P_1–N (or P_2–N or P_3–N) voltage of the star system. Switching between star/delta operation can figure in the starting sequence of large motors. Final circuit supply voltages of P_1–P_2 or P_1–N, etc. along a street give flexibility to supply systems and their loading.

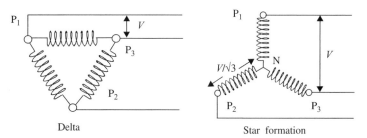

Delta   Star formation

demagnetising factor When a magnetic circuit involves a sudden change of cross-section or material permeability, fields emerge which tend to oppose the prevailing magnetisation, e.g. an infinitely long rod has no demagnetising effects. A very long rod has only a small demagnetising field at its ends.

A flat plate produces a powerful demagnetising field and is very difficult to magnetise. Tables of the demagnetising factors (and relevant approximations) are published, but today computer methods allow for exact numerical solutions fairly easily.

die   This is the apertured metal block into which a punch is forced to cut out a shape from offered metal sheet.

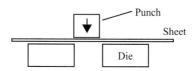

Materials used are alloy steel and tungsten carbide.

dislocation   Metals form into grains, each of which has a regular crystal lattice. Defects in the perfection of the lattice are called dislocations. Dislocations are created by cold work, radiation damage, foreign atoms, etc. Under the influence of heat and stress, dislocations may move through the lattice. Annealing can be done in such a way that most dislocations are swept out.

dynamic friction   Once two contacting surfaces are in motion with respect to each other a frictional force appears between them opposing the sliding action. This force is proportional to the pressure of contact. Such sliding friction may have a different value from static friction which is evident up to the point where relative motion occurs. The automated handling of steel sheet benefits from a low dynamic friction whereas the stability of large cores built of many plates depends on a minimum value of static friction being present.

dynamo   This is a direct current generator in which an arrangement of switches (commutator) ensures that the emf generated in rotating coils contributes to a unidirectional output current (DC).

Earnshaw's theorem   This theorem relates to the stability of statically poised systems and suggests that a body cannot be in a stable position unless at least one degree of freedom is removed. Special magnetic arrangements such as servomechanisms, field gradient control and superconductors can be set up to make it appear that Earnshaw's theorem is being violated.

eddy current   When a time-varying magnetic field encounters a conductive body the emfs generated lead to current loops within the body. Very often eddy currents are undesired and methods to reduce them are applied, such as the lamination of transformer cores or the transposition of copper conductors. In the case of transverse flux heaters and eddy current melting, the eddy currents are the desired result.

edge drop   This is the inverse of crown, and relates to the reduction in thickness of rolled steel sheet which occurs near the edges of sheet. It is an undesired effect as stacks of laminations which include some of varying thickness across their width can assume an unwanted curved shape. Edge drop can be reduced by the use of specially profiled rolls along with the setting of hot mill rolls at a slight 'cross angle'.

exothermic gas   When fuel gas is burnt with air the products of combustion can be tailored to fit a variety of steel annealing treatments. The range of air/gas feedstocks employed is such that thermal energy is released during reaction.

extension passing   (temper passing) Steel strip may be given a light rolling pass which causes a length increase of a few per cent. This treatment tailors the physical hardness of strip and prepares the way for critical grain growth on subsequent anneal. At the small percentage extension used (e.g. 8 per cent) the percentage reduction in thickness is numerically very similar.

farad   Unit of electrostatic capacity as applied to capacitors (condensers). A capacity of 1 farad stores one coulomb of electricity for each volt of potential sustained, i.e. $Q = CV$, Q is the charge in coulombs, V is the voltage, C is the capacity in farads. Energy stored is $E = \frac{1}{2}CV^2$, where E is energy in joules.

Faraday cage   A system of linked conductors enclosing a space so as to shield it from electric fields.

Faraday's laws of induction   When the flux linking a conductor system varies with time then the induced emf in the conductor system is $V = -\mathrm{d}\phi/\mathrm{d}T$, where $T =$ time, $\phi =$ magnetic flux. The negative sign expresses Lenz's law so that the field arising from any current which flows in the conductor system opposes that producing the emf.

Fleming's right and left hand rules   These rules express the mutual orthogonality of magnetic field, motion and current direction. The right hand rule applies to generation of current and the left hand rule to motor effects.

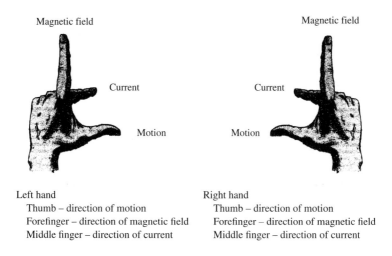

Magnetic field   Magnetic field

Current   Current

Motion   Motion

Left hand   Right hand
Thumb – direction of motion   Thumb – direction of motion
Forefinger – direction of magnetic field   Forefinger – direction of magnetic field
Middle finger – direction of current   Middle finger – direction of current

four-terminal resistor   When current is to be measured by observing the voltage dropped across a resistor it is convenient to feed current in and out near the ends of the resistor, but to measure volt drop across a shorter in-board portion with terminals away from the current feed points. In this way 'contamination' of the volt drop by the current patterns near the feed points is avoided.

fully finished   or fully processed,   Steel requiring no further heat treatment before use.

gamma rays   These are electromagnetic radiation of very short wavelength arising from nuclear decay in an unstable isotope. Gamma rays are used to penetrate steel and form the basis of thickness-measuring devices. A typical isotope used is americium 241.

grains   Metals crystallise in the form of grains, each of which is composed of an orderly lattice of atoms. Many metallurgical operations modify the grain size, structure and orientation as part of routes to development of favourable magnetic and other physical properties.

H coil   In magnetic measurement the magnitude of an applied field (H) can often be inferred from the size of the current in the conductor system producing it. However it is sometimes convenient to place small air-cored coils in positions where an exact field assessment is required. In AC systems the emf produced in an H coil is proportional to dH/dt and has to be integrated to give an H signal. However $\hat{H}$ is relatable to the average value of the H coil output i.e. $\hat{H} \propto \overline{|V|}$.

Helmholtz coils   A system of coil pairs set up so that an accessible space between them experiences a uniform magnetic field over a distance large enough to conduct experimental work within it.

henry   The henry is a unit of inductance. An inductance (system of conductors around which a magnetic field is formed when current flows) of one henry is such that if the current through it changes at the rate of 1 volt/second an emf of 1 volt will be induced in it:

$$V = -L\frac{dI}{dt}$$

where V is in volts, L is in henries, I is current, and t is time. The minus sign indicates conformity with Lenz's law.

hot band   Steel from the hot strip mill which will not be made thinner by further hot rolling, but will go onwards to cold rolling.

interpoles   When current flows in the armature of a DC motor the magnetic field arising distorts the stator field pattern. Extra stator poles placed between the main stator poles can be fed with some armature current and offset this effect. Alternatively commutator brush positions can be altered to maintain low spark commutators.

joule   Unit of energy. 1 joule $= 1$ watt second.

leader ends   In order to preserve the ends of special-quality steel coils from the hazards of coil joining, rolling completion, etc. it is sometimes the practice to weld lengths of inexpensive expendable steel onto the coil ends so that top-quality material is not wasted.

leakage inductance   When, in transformers, some of the flux arising from current change in one winding does not link with a second winding, the flux not linking behaves as would a simple inductor lacking a secondary to link with. This effect presents itself as the transformer primary winding having an inductance independent of that happening at the secondary. It can lead to poor regulation (q.v.) which may be a detriment or a useful feature depending on the application.

lean alloys Silicon and other elements are often added to electrical steels to secure particular effects. When it is desired to maintain the maximum high-field permeability of steels, alloying is minimised. Lean alloys may be considered to be iron with less than 0.2% of added elements.

Lenz's law Although Lenz's law is usually quoted in relation to the sign of induced currents the minus sign which arises really expresses the law of conservation of energy which would be violated if cutting magnetic field lines with a conductor could lead to a current able to augment the primary motion.

Litz (Litzendraht) At high frequencies the skin effect causes current to flow mainly in the surface layers of a conductor leaving the interior unused. This increases the high-frequency resistance of conductors. Wire made up of a rope of many strands each insulated from the others uses the copper involved much more efficiently. Such Litz wire is used in high-frequency applications where kiloherz operation applies.

lumen This is a unit of light output. 4π lumens are emitted by a source of one standard candle power. The efficiency of an electric light source is quoted in terms of lumens per watt of input power.

magnetic field This is the region within which magnetic effects can be detected. Deflection of a compass needle, production of an emf in a rotated coil, etc. reveal the presence of a magnetic field. Magnetic fields arise from electrons in motion either in the form of current flow or the spin of electrons in a crystal lattice.

Magnetic fields are detectable at a distance from the charge movements responsible but the mechanism of field propagation and maintenance is obscure.

magnetostriction Magnetostriction is the change in length of a solid in the direction of magnetisation of solid magnetic material when magnetised. It can be negative or positive and is of the order of parts per million. Magnetostriction is independent of the polarity of the magnetisation and occurs at twice the magnetising frequency.

mean or average The average value of the emf generated in a B or H coil is proportional to the peak value of B or H.

mega A scaling factor $= \times 10^6$; examples are megavolt, megapascal (pressure), etc.

mesh See delta/star. A delta configuration is often called a mesh.

microstrain Describes a change in dimension of 1 part per million of normal length. Changes in dimensions to which this is relevant are magnetostriction, thermal expansion, elastic deformation under stress, etc.

mmf, magnetomotive force The magnetising force H (in A/m) which when applied yields a magnetisation B in tesla.

mon A unit of 'shape' in which the length difference of adjacent regions is 0.01%.

MPa, megapascals   A pressure of 1 pascal arises from the application of a force of 1 newton over an area of 1 square metre. The pascal is a very small unit for practical use and the megapascal is more convenient.

mutual inductance   If two circuits are close to each other then changing current in one of them can induce an emf in the other. The emf induced in the second circuit V is proportional to the rate of change of flux (for air cored systems, of current) in the first circuit such that $V = -M \, dI/dt$. M is quoted in henries.

Newton force   A force of 1 newton applied to a mass of 1 kg will cause it to accelerate at a rate of 1 metre/second/second.

Newton's apple   Traditionally Newton was caused to ponder the presence of gravity by noting the fall of an apple from a tree. He speculated on why the apple should move and postulated a force to engender motion. Gravity, like magnetism, is propagated by means which remain obscure. It may be noted that a small apple of 1/10 kg mass (approx $\frac{1}{4}$ lb) rests on the hand with a force due to gravity (on Earth) of about 1 newton.

negative feedback   In any process in which a desired outcome (offered as a pattern) is compared with an actual outcome the difference can be fed back to the input so that the undesired deviation is diminished. In amplifiers this enables a sinusoidal waveform of magnetisation to be enforced upon a magnetic material. This condition is often required to enable magnetic materials to be tested under specified sinusoidal conditions. The application of negative feedback diminishes the gain of a system, and if overused runs the risk of producing instability. However computer-controlled digital methods of feedback and waveform control offer the prospect of controllable waveform-correcting amplifiers. The judicious application of negative feedback covers up the internal non-linearities of amplifiers and reduces costs.

ohm   The unit of electrical resistance. Ohm's law indicates that a current of 1 ampère flowing through a resistance of 1 ohm experiences a potential drop of 1 volt. The ohm has been expressed as the resistance of a column of mercury 106.3 cm long of 1 sq mm cross-section at $0\,^\circ$C. Modern definitions appeal to more absolute methods. The ohm is also the unit used for reactance and impedance in AC circuits where the ratio of voltage to current is quoted in ohms but where the reactance or impedance contains a mixture of reactive and resistive items.

pascal   See MPa, megapascal.

path length   Where a magnetic circuit is made up of inhomogeneous components such as a strip of steel and a flux closure yoke the exact length of steel sample considered to be involved in, for instance, power loss measurement, may become difficult to define. The path taken by flux varies with the intensity of magnetisation of samples and its response to air gaps. Efforts have been made to standardise path lengths by allocating 'agreed' values to specific geometries of measurement systems.

peak   This is the maximum value of any AC quantity. For sinusoidal currents or voltages the peak value is 1.414 times the RMS (root mean square) value, i.e.

$$\hat{V} = \sqrt{2}\tilde{V} \quad \text{where } \hat{V} = V_{\text{peak}}, \ \tilde{V} = V_{\text{RMS}}, \ 1.414 = \sqrt{2}$$

Further, the average value of a sine wave relates to the RMS value thus:

$$\frac{\text{RMS}}{\text{Average}} = 1.111 \qquad \text{Average} = \frac{\text{RMS}}{1.111} = \text{RMS} \times 0.9$$

So

$$\hat{V} = \text{RMS} \times 1.414, \quad \overline{|V|} = \text{RMS} \times 0.9$$

permanent magnet   A magnetised body whose internal domain structure remains arranged so that a considerable field is produced outside it in the absence of an external magnetising force. To qualify as 'permanent' its magnetism should be resistant to some degree of vibration and moderate demagnetising fields.

phosphate coating   Grain-oriented steel carries a silicate glass coating developed during high-temperature annealing. This is usually supplemented by an additional phosphate (plus other proprietary compounds) coating to improve tensile effects, etc. Non-oriented steels expected to operate at high temperatures can be given a phosphate insulating layer which performs well in this rôle but has adverse wear effects on punching tools.

piezo-electric   When certain crystals are exposed to stress in particular directions they develop electric charges on their surface. These charges can be measured and related to stress values. Such piezo-electric crystals are incorporated into accelerometers for magnetostriction measurement, gramophone pick-ups, microphones, etc. Materials used include barium titanate, quartz, etc.

power factor   When an electrical load is purely resistive it dissipates power in line with the rate $V \times I$, where V is the voltage input and I the current drawn. When the load is supplied with alternating current and contains reactive components (inductance, capacity) the product $V \times I$ can greatly exceed the actual dissipation of power in the load. A pure capacitor would 'borrow' and pay back energy from a supply each cycle without actual dissipation of energy but the current flow in and out of it could be considerable.

Transformers, inductors, induction motors and the like often operate at a power factor below 1.0 (1.0 for a pure resistor, 0.0 for a perfect capacitor or inductor). Low power factors are undesired since the current supply arrangements have to cope with the flows involved in borrowing and paying back energy cycle by cycle as well as useful current flow concerned with actually dissipating heat, mechanical effort, etc. Various strategies exist to improve the power factor, notably the placing of capacitors in parallel with inductive loads.

The rating of a device in VAs (volt–amp product) is indicative of the current likely to be drawn, and not of the watts likely to be dissipated which may be a lot lower. Further, a device rated in watts may draw much higher currents than watts/V may imply if it operates at a low power factor.

Power factor could be considered as the ratio of

$$\frac{\text{useful or dissipative work}}{\text{apparent work given as a VI product}}$$

Values below 0.7 are unwelcome in supply systems and 0.95 and above is preferred.

Power factor $= \cos \theta$ where θ is the angle between voltage and current. When $\theta = 0$, $\cos \theta = 1.0$.

punch/die   See die.

Q Q is the quality factor of a circuit or system. It is usually represented by the ratio of the energy contained in a vibrating system to the energy dissipated per cycle of operation. Where losses in a system are high, Q values must remain low. High Q systems at resonance show a sharp high amplitude peak of voltage or movement.

Mechanical systems of high Q are bells. A low Q system is a sheet of lead. If dropped it produces a dull thud not a sustained note. It may be noted that a lead bell cooled to liquid nitrogen temperatures produces a good ringing tone until warmed up.

RCP, Rogowski–Chattock potentiometer   This is a device for detecting differences in magnetic potential from place to place. This ability is valuable in the devising of magnetising systems able to produce regions of uniform magnetisation for measurement purposes. The principles involved are well described in an article in *Archiv für Electrotechnik*, **1**, 4 (1912), pp. 141–51, by W. Rogowski and W. Steinhaus.

regulation   When a load is placed on a power supply system on nominal voltage V, the load current will cause the output voltage to be depressed. The degree of this fall is called regulation. A supply able to maintain 98% of nominal voltage at full rated current supply would be considered to have 2% regulation (very good). A fall to 70% would be poor. When domestic water taps are opened the supply pressure often falls drastically during the period of outflow.

RMS   This is the root mean squared value of usually voltage or current and relates to that value which represents the effective or heat producing quality of the supply; see peak and average.

A DC supply of 100 volts would, at 10 amps, deliver 1000 watts (1 kW) in a $10\,\Omega$ resistive load. An alternating supply connected to the same load may alternate between $+141.4$ and -141.4 volts but produce the same heat output as a 100 volt DC supply. So 100 volts is the declared effective value of the AC supply.

Note $100 \times \sqrt{2} = 141.4$

rolling direction   Steel strip is produced in hot and cold rolling mills by rolling the flat strip thinner at each rolling operation. The width of strip remains substantially the same, its thickness is reduced and its length increases. The direction of length increase is called the 'rolling direction' and is often quoted or referred to as a reference direction to which directional variation of magnetic properties may be referred.

saturation   When an increasing magnetising field is applied to a ferromagnetic material its magnetisation increases rapidly at first, then more slowly until all the available domain rearrangement and vector rotation processes have been exhausted. This last state is termed magnetic saturation. 'Technical saturation' is considered to have been reached when the metal has come within a few per cent of ultimate saturation.

scroll slitting   Slitting to give a wavy edged slit. This reduces waste for the stamping of round laminations.

semi-finished   Some electrical steel is delivered with its full magnetic potential developed, whereas other steels are supplied in a condition which is optimal for punching by UK/USA methods but requires a final heat treatment to develop optimum magnetic properties. Such steels are termed semi-finished or semi-processed.

shaded pole   If a copper or aluminium ring of low resistance is placed round a batch of steel laminations the time progress of flux within that core is delayed by the opposing effect of eddy currents flowing in the 'shading ring'. This opposition is a response to Lenz's law. Such a phase delay of magnetisation can be exploited to help create a rotating field in small motors when it is combined with undelayed flux from an unshaded part of a core.

shape   Many factors combine in a strip rolling process to cause strip to have some of the notional filaments of its composition longer than others. Thus strip can have edges shorter than its centre regions or vice versa. Strip with longer middle filaments is said to have a 'loose middle' and 'tight edge'. Strip with short central filaments will have a wavy edge. A complex system of assessment and control of these manifestations of shape is practised so that efficiency and convenience of processing is preserved during all rolling operations. (See mon.)

silicate glass   The magnesium oxide coating applied to grain-oriented steel reacts with the silicon content of the steel to produce a magnesium silicate 'glass' (in fact consisting of tiny crystals) coating.

skin effect   See Litz.

slip   In an induction motor delivering power to a load there is a small difference between the rpm of the machine rotor and the rotating field produced by the stator poles. This difference is known as 'slip' and is at a level of a few percent over the working range of the machine. Off-load, slip reduces to a tiny amount so that just the losses in the machine are offset within a limited range. Growth of slip leads to growth of power output.

slitting   Electrical steel strip as produced (e.g. 1 m wide) may be too wide to suit stamping and core cutting machines. A slitting operation cuts wide coils into narrower ones. Essentially steel strip is fed through sets of circular slitting blades, which may be either powered or passive (pull through slitter), which slice the strip into narrow portions which are rewound into narrow coils.

Slitting is possible with steel blades, but tungsten carbide types are preferred. The setting up and operating of slitters is a very complex skill and great care and attention must be paid to the process to yield high-quality burr-free strip. Computer methods may be used to calculate blade set-up assemblies and to optimise the width which can be extracted from wide strip so that scrap loss is minimised.

squareness   Highly permeable steels arrayed in an efficient magnetic circuit are able to switch a large percentage of their $-B_{sat}$ to $+B_{sat}$ flux with the application of a very small magnetomotive force. The BH loops concerned have near-vertical portions and high square shoulders. Various numerical expressions of squareness can be used. This property is important for the design of pulse transformers.

squirrel   It is necessary from time to time to monitor the conditions applying in strand anneal furnaces and quasi-batch tray pushing furnaces. The use of trailing thermocouples is very hazardous and wasteful. Devices have been developed in which a heavily insulated box containing a data recorder is passed through the furnace coupled to short thermocouples monitoring the box environment. Such a device is called a 'squirrel' and its shape and time–temperature endurance is tailored to the duty involved. (Note the use of 'pigs' in oil pipelines. See also Chapter 4 for types.)

star/delta   See delta and mesh.

static friction   See dynamic friction.

superconductivity   Some metals and compounds lose all ohmic resistance at low temperatures. This effect is known as superconductivity. There is a limit to the currents which may be carried as the resulting magnetic field lowers the maximum temperature at which superconductivity can be sustained.

temper rolling   See extension passing. A light cold rolling operation applied to steel strip after a grain-growing and decarburising anneal.

tesla   Unit of magnetic flux density, see weber.

time constant   When a process proceeds at a rate proportional to the amount of the process remaining uncompleted, such as the discharge of a capacitor via a resistor, the process is exponential in form and notionally never complete. The initial rate of discharge, if sustained, would empty the capacitor after a time known as the time constant. The index of the controlling exponential is multiplied by CR for capacitance systems and L/R for inductors. Units are henries, farads and ohms for scaling in seconds.

torque   This is the turning effort developed by a motor shaft, or applied to a generator, and is a product of force (newtons) × radius of applications (metres). Torque is quoted in newton metres.

universal motor   This is a wound armature commutator machine designed so that it will operate with acceptable efficiency on an AC or DC supply. These are usually of the series operation type and are used for vacuum cleaners, food mixers, etc.

vacuum degassing   This is part of the steel-making process in which the metal is exposed to vacuum and maybe injected oxygen or inert bubbling gas to assist in the removal of carbon. It leads to a very clean (inclusion-free) steel.

virtual instrument   This is an adaptation to a computer which enables it to simulate the functions of a digital voltmeter as well as many other instruments. By changing software the instrument type is altered at once.

volt   A unit of potential difference. If 1 coulomb of electricity is moved through a potential difference of 1 volt then 1 joule of work is expended. The voltage of a Weston standard cell is close to 1.0188 volts.

watt   Unit of power (rate of doing work). One watt = one joule/second, or one amp falling through one volt. Note: watts $= V \times I = I^2 R = V^2/R$.

wattmeter   This is a device for indicating the rate of power dissipation in a circuit. It is designed to indicate real power, i.e. $V \times I \cos \theta$ rather than simply $V \times I$. It is applied to the testing of magnetic materials.

weber   Unit of quantity of magnetic flux. If the flux contained by a closed circuit varies at the rate of 1 weber/second then 1 volt will be induced in that circuit. Note: the tesla is a unit of flux density, so that 1 weber is contained in an area of one square metre if the intensity of magnetisation is 1 tesla, i.e. webers = teslas $\times$ area.

yoke   When a magnetic circuit would otherwise contain large air gaps a ferromagnetic yoke can be applied to bridge these gaps and render the circuit more magnetically efficient.

Young's modulus (E)   This is an indicator of by how much a material will deform under stress. Normally it is quoted as the ratio stress/strain:

$$\text{Stress} = \frac{\text{force}}{\text{area}} = \frac{F}{A}, \quad \text{Strain} = \frac{\text{length change}}{\text{length}} = \frac{\Delta l}{l}, \quad \therefore E = \frac{Fl}{A \Delta l}$$

Appendix

Conversion factors

Tesla, other units	1 gauss $= 10^{-4}$ tesla
	1 kilogauss $= 10^{-1}$ tesla
A/metre, other units	1 oersted $= 79.58$ A/m
Kilowatt (1000 watts)	1 horsepower $= 0.746$ kW
Watts/kg	1 watt/kg $= 0.4536$ watts/lb
Webers	1 weber/square m $= 1$ tesla

Lines per square cm $=$ gauss $= 10^{-4}$ tesla

Lines per square inch (USA) $= 0.1555$ lines/square cm

Note: in permanent magnet work and in the USA considerable use is still made of the gauss and oersted. In the gauss/oersted system the permeability of free space is 1.0 , while in the tesla, A/m system it is $4\pi \times 10^{-7}$.

Square inches (USA) $=$ square centimetres $\times 0.155$

Square centimetres $=$ square inches $\times 6.4516$

1 inch $= 2.54$ centimetres

1 metre $= 39.37$ inches

1 kg $= 2.2046$ lb

1 lb $= 0.45359$ kg

1 lb/square in $= 6.895$ kPa

1 ton/square inch $= 15.44$ MPa

Density of low alloy steel $= 7.86$ g/cm^3

Appendix

Useful formulae

$$|\overline{V}| = 4\hat{B}nf A \qquad \text{Transformer equation}$$
$$\tilde{V} = 4.44\hat{B}nf A \qquad \text{Transformer equation}$$
$$V_{PK} = V_{RMS} \times 1.414$$
$$V = IR \qquad \text{Ohm's Law}$$
$$W = I^2 R = V^2/R = V \times I \qquad \text{Power equation}$$
$$X_L = \omega L, \quad \omega = 2\pi f \qquad \text{Inductive reactance}$$
$$X_C = 1/\omega C \qquad \text{Capacitive reactance}$$
$$Z = \sqrt{R^2 + X_C^2 + X_L^2} \qquad \text{Impedance}$$
$$PF = \text{watts}/(VA), \quad PF = V \times I \cos\theta$$
$$V_{ph-ph} = V_{ph-N} \times \sqrt{3}$$
$$I = dQ/dt$$

Time constant $= CR$ or L/R

Resistors in parallel $1/R_{Tot} = 1/R_1 + 1/R_2 + \cdots$

Resistors in series $R_{Tot} = R_1 + R_2 + R_3 + \cdots$

Inductors in series $L_{Tot} = L_1 + L_2 + L_3 + \cdots$

Inductors in parallel $1/L_{Tot} = 1/L_1 + 1/L_2 + 1/L_3 + \cdots$

Capacitors in series $1/C_{Tot} = 1/C_1 + 1/C_2 + 1/C_3 + \cdots$

Capacitors in parallel $C_{Tot} = C_1 + C_2 + C_3 + \cdots$ Reactance of capacitor $= 1/(\omega C)$

Reactance of inductor $= \omega L$

Force between magnets $\propto B^2 A$

Peak induction $\hat{B} = |\overline{V}|/(4nf a)$

Example of use of transformer equation: $\tilde{V} = 4.44\hat{B}nf A$.

Given a core of cross-section $0.01\,\text{m}^2$ and a frequency of operation of 50 Hz working at $\hat{B} = 1.5\,\text{T}$, calculate the volts per turn arising:

$$\tilde{V} = 4.44 \times 1.5 \times 1 \times 50 \times 0.01 = 3.33\,\text{volts}$$

Note: Where the stacking factor may be less than 1, e.g. 0.97, adjustment must be made to the result. If 1.5 T applies to the core cross-section including 'air frame' the actual induction in the strips will be $\frac{1.5}{0.97}\,\text{T} = 1.546\,\text{T}$.

Derivation of relationship between average volts and peak induction

$$\overline{|V|} = 4fNA\hat{B}, \quad \text{and} \quad e = \frac{d\Phi}{dt} = NA\frac{dB}{dt}$$

so

$$e\,dt = NA\,dB$$

Integrating both sides

$$\int e\,dt = NA\int_{-\hat{B}}^{+\hat{B}} dB = 2NA\hat{B}$$

But $\bar{V} = \int \dfrac{e\,dt}{T/2}, T = \dfrac{1}{f}, \dfrac{T}{2} = \dfrac{1}{2f}$, but since

$$\int e\,dt = 2NA\hat{B}$$
$$\bar{V} = \int e\,dt.2f$$
$$\bar{V} = 2NA\hat{B}.2f$$
$$= 4NAf\hat{B}, \quad V \text{ volts, } B \text{ tesla, } f \text{ Hz, } A \text{ m}^2$$

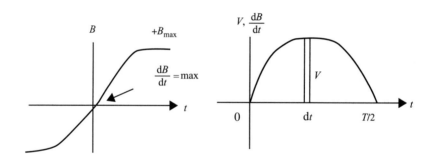

List of symbols

ω	$2\pi f$
μ	Permeability, tesla/A/m
μ_0	Permeativity of free space, $4\pi \times 10^{-7}$
μ_r	Relative permeability μ/μ_0
μ_{max}	Maximum permeability, tesla/A/m
μ_i	Initial permeativity, tesla/A/m
Φ	Quantity of flux, webers
ρ	Resistivity, $\Omega\,m$, ohm metres
A	Area, m^2 (sometimes a)
B	Induction, tesla
$\hat{B}$	Peak induction, tesla
B_{rem}	Remanent magnetism, tesla
B_{sat}	Saturation induction, tesla
C	Capacity, farads
f	Frequency, Hz
H	Applied field, A/m
$\hat{H}$	Peak applied field, A/m
H_C	Coercive force, A/m
I	Current, amps
J	Intrinsic magnetisation ($B-H$), tesla
J_{sat}	Saturation intrinsic magnetisation, tesla
L	Inductance, henries
n	Number of winding turns (sometimes N)
R	Resistance, ohms
t	Time, sometimes T
V	Voltage, volts
$\bar{V}$	Average voltage, volts
$\overline{|V|}$	Average rectified voltage, volts

$\tilde{V}$ RMS (effective) voltage, volts

$\hat{V}$ Peak voltage, volts

W Power, watts

X_C Capacitive reactance, ohms

X_L Inductive reactance, ohms

Bibliography

BAILEY, A. R.: 'A textbook of metallurgy' (Macmillan, London, 1961).

BOZORTH, R. M.: 'Ferromagnetism' (Van Nostrand, first edn 1951).

BRAILSFORD, F.: 'Physical principles of magnetism' (Van Nostrand, 1966).

CONNELLY, F. C.: 'Transformers' (Pitman, London, first edn 1950).

COTTON, H.: 'Applied electricity' (Cleaver Hume 1951). Various books by Cotton cover the subject in greater depth.

DUFFIN, W. J.: 'Electricity and magnetism' (McGraw-Hill, 1990).

GOLDING, E. W.: 'Electrical measurements and measuring instruments' (Pitman, London, 1949).

HEATHCOTE, M.: 'J and P transformer book' (Newnes, Oxford, 1998).

JILES, D.: 'Magnetism and magnetic materials' (Chapman & Hall, New York, London, 1991).

KARSAI, K., KERENYI, D. and KISS, L.: 'Large power transformers' (Elsevier, Amsterdam, 1987).

KAY, G. W. C. and LABY, T. H.: 'Physical and chemical constants' (Longmans, London). Many revisions; an old edition and an up-to-date edition complement each other.

McGANNON, H. E. (Ed.): 'The making, shaping and heat treatment of steel' (US Steel, first edn 1964).

Index